Neuroscience Made Easy

Jon Adams

Copyright © 2024 Jonathan Adams

All rights reserved.

ISBN: 9798321327876

CONTENTS

1 Neurons The Building Blocks .... Pg 6

2 The Electric Brain Neural Communication .... Pg 19

3 Brain Architecture Organization and Function .... Pg 30

4 Sensory Systems and Perception .... Pg 46

5 Memory and Learning .... Pg 56

6 Consciousness and Cognition .... Pg 68

7 Emotions and the Social Brain .... Pg 87

8 Neuroplasticity The Adaptable Brain .... Pg 96

INTRODUCTION

Welcome to "Neuroscience Made Easy," a journey into the intricacies of the human brain that doesn't require a GPS or a PhD to navigate. This book is your personal guide through the meandering pathways of neurons, synapses, and the remarkable processes that define every thought, memory, and emotion carving the human experience.

Stepping into the world of neuroscience might seem as daunting as scaling a mountain, but fear not—this book aims to turn the steep cliffs into manageable trails. Here, complex concepts are broken down into digestible pieces, removing the cloak of technical jargon and revealing the marvels of the mind in plain language.

As you turn these pages, you'll unearth the secrets behind the brain's incredible plasticity, you'll understand what lights up the brain when you fall in love, and you'll have a front-row seat to the symphony of neurotransmitters that enable you to savor a delicious meal or cherish a sunset.

"Neuroscience Made Easy" places significant emphasis on not just conveying information, but on why this information matters to you. It's a spotlight on the brain's role in shaping behavior, forming habits, and its profound impacts on health, well-being, and the richness of life.

Whether you're a student keen on understanding the basics for the first time, a teacher looking for a valuable resource, or a life-long learner with an insatiable curiosity, this book is for you. It's peppered with relatable examples, everyday analogies, and engaging explanations that make neuroscience not only approachable but downright intriguing.

Embark on this enlightening expedition to not just discover, but genuinely appreciate the masterful workings of the most complex organ in the human body—the brain. Welcome to a ride through neuroscience that is as enjoyable as it is educational, where every new fact is a piece of the puzzle that makes us who we are.

NEURONS THE BUILDING BLOCKS

Welcome! You're about to dive into the intricate world of neurons—think of them as the diligent workers in the brain's bustling metropolis. Each neuron, like an electrician, tirelessly ensures that the signals guiding your thoughts, feelings, and movements are relayed correctly. Imagine each neuron as a tiny worker in your brain's command center, wiring thoughts and reflexes into action. As these industrious cells send electrical signals, they're using a language of chemistry as complex and nuanced as any spoken word. To make sense of this, picture something as everyday as a city's traffic lights; they flash signals to orchestrate the flow of countless vehicles. Neurons do something similar by rapidly firing signals that manage your body's actions and reactions. Where the plot thickens, and things get really intricate—like understanding the exact play-by-play of a signal's journey from one neuron to another—think of a relay race with each runner passing the baton with precision.

Your brain's ability to learn a new skill, say playing the guitar or mastering a new language, is thanks to something called neuroplasticity. It's much like a path in a forest that becomes clearer and easier to walk the more it's traveled. If this all sounds like navigating a labyrinth, you're right, but fear not—the roadmap we're about to follow is sprinkled with analogies and visuals to keep things crystal clear. By the end, you'll see that each detail, each tiny worker in your brain, plays a vital role in crafting the rich tapestry of daily life, and you'll appreciate the genius behind nature's design. So, let's step into this amazing cerebral architecture and illuminate each corner with the bright light of understanding.

Think of neurons as the specialized cells that act like the brain's communication experts. They are tiny units, yet they pack a powerful punch when it comes to running the show that is the nervous system. Imagine a complex network of cables running all throughout a supercomputer; neurons in your body are quite similar. They form a vast network that sends signals at lightning speed to keep everything from your heart beating to remembering your best friend's birthday.

Neurons are considered the building blocks because they are the primary cells that make up the nervous system and execute all its functions. Just as the foundation of a house determines the sturdiness of the entire structure, neurons lay the groundwork for the brain's operation and the body's responsiveness. Each neuron is outfitted with three critical parts: the dendrites, cell body, and axon. Dendrites pick up incoming signals, the cell body processes those signals, and the axon sends out the orders to other neurons, muscles, or glands. This process is akin to a busy office receiving mail, sorting through the information, and forwarding instructions to the right departments.

Analogies aside, neurons communicate through a mix of electricity and chemicals. Imagine texting your friend, the electric signal is your typed message moving through your phone, and the notification your friend hears is the chemical part of the process as their phone translates that signal into sound. It's this sophisticated system that enables us to interact with the world in real-time, understand complex ideas, and create memories.

The mechanics of these neurons can be intricate, yes, but at heart, they're about connection and communication, something inherently familiar to every one of us. These little biological marvels show both technology's sophistication and its potential for growth. Through this lens, neurons aren't just biological cells; they're the equivalent of superhigh-tech devices enabling your body to function as a well-oiled machine, with a dab of innovation keeping things running smoothly and tirelessly working to learn and adapt. This insight is not an endpoint but rather a beginning to understanding the impressive silent symphony your body conducts every moment to keep you alive and interacting with your surroundings.

A neuron at rest is like a tiny battery, not doing much until it gets the right signal. It's sitting there with a resting membrane potential, which is a bit like the charge on that inactive battery; this charge is negatively charged compared to the outside. This is because of different concentrations of ions, like sodium and potassium, on each side of the neuron's cell membrane.

When a stimulus kicks in, it changes that resting potential. If it's strong enough, it leads to depolarization. This is where the cell's inside becomes less negative compared to the outside, thanks to positively charged sodium ions rushing in through special protein channels.

If this change is significant, it can reach what's known as a threshold, and that's where things get exciting. Reaching this threshold triggers the action potential, the neuron's way of passing on a signal. During an action potential, a rapid rise and fall in electrical charge along the neuron's axon occurs.

After the rapid influx of sodium ions, the neuron needs to reset itself. That's called repolarization, where potassium ions move out of the cell, which helps bring the inside of the cell back to its negatively charged resting state. Sometimes the charge overshoots a bit, becoming even more negative than at rest, which is known as hyperpolarization. This is like an overcorrection, but the neuron has ways (pumping the ions back to right places) to bring things back to the resting potential.

But the story doesn't end at repolarization. The signal is still on the move. At the end of the axon are the synaptic terminals, waiting to pass the message on. When the action potential reaches these terminals, it signals little vesicles packed with chemicals called neurotransmitters to merge with the terminal membrane, spilling their content into the synaptic cleft, which is the gap between one neuron and the next.

These neurotransmitters then attach to receptors on the next neuron's dendrites. Think of them as keys (neurotransmitters) fitting into locks (receptors), and when they fit, they cause changes in the next neuron, passing along the message that can eventually trigger another action potential there.

This entire process is how neurons communicate, and each step is crucial to ensure the message gets through. Understanding these technical details demystifies how thoughts translate to action or how sensations become perceptions. It's a lot to take in, but each concept provides a critical piece of the puzzle for how humans experience the world and react to it.

Imagine you're in a bustling city; that city is your brain, and the neurons are like the city's residents. The dendrites of a neuron are like the antennae on a house, picking up signals and messages from the outside, which translate to the all sorts of information gathered throughout the day. Then there's the neuron's cell body—it's the central hub, like a post office, sorting the mail, which in this case is the information received, deciding what's junk and what's important enough to pass on.

Next comes the axon, think of it as the highway where the sorted mail, now crucial messages, are transported swiftly across town. This highway is not for everyone; it's exclusive for vital information that dictates actions and decisions. At the end of this highway, you hit the synaptic terminal, much like a distribution center, which decides the final destination of these messages. Here, little packets of neurotransmitters are neatly packaged and readied for delivery across the synaptic cleft, akin to the gap between one island to another.

This gap is bridged when the neurotransmitters are released from the synaptic terminal, soaring across the synapse like a fleet of drones carrying packages to drop on the doorstep of the next neuron's dendrites. These deliveries are essential because they determine whether the receiving neuron will spring into action or stay at rest, influencing everything from the beat of your heart to your ability to remember the taste of your favorite coffee.

And why does this matter? Well, just as every task in a city keeps the lives of its residents moving smoothly, every part of a neuron plays a crucial role in guiding how you experience and respond to the world. It's a miraculous daisy chain of events that happen in mere milliseconds, yet they impact every facet of your life. This fascinating interplay within your nervous system illustrates the elegance with which your body operates, echoing the vibrancy and efficiency of a city teeming with life.

Here's a step-by-step account of how signals flow through the neuron's synaptic terminal:

- Synaptic Terminal Operation

  - Action Potential Arrival:

    - When an action potential reaches the synaptic terminal, it prompts calcium ion channels to open.

    - Calcium ions flow into the cell, creating an environment that encourages vesicles to move toward the terminal membrane.

  - Vesicle Fusion:

    - Synaptic vesicles, which contain neurotransmitters, are directed to the membrane.

    - The influx of calcium ions triggers a reaction that allows vesicles to fuse with the terminal membrane.

    - This fusion results in the release of neurotransmitters into the synaptic cleft.

- Types of Neurotransmitters and Functions

  - Acetylcholine:

    - Often involved in muscle movement and memory formation.

  - Dopamine:

- Plays a key role in reward, motivation, and motor control.

- Serotonin:

  - Mostly associated with mood regulation, appetite, and sleep.

- GABA:

  - Acts as a primary inhibitor in the brain, preventing overexcitement.

- Glutamate:

  - The most common excitatory neurotransmitter, involved in learning and memory.

- Receiving Mechanisms

  - Neurotransmitter Diffusion:

    - Released neurotransmitters diffuse across the synaptic cleft.

  - Receptor Binding:

    - Neurotransmitters bind to specific receptors on the post-synaptic neuron's dendrites.

    - This binding can either initiate or prevent an action potential, depending on the neurotransmitter type.

  - Excitatory and Inhibitory Impact

- Excitation:

- Excitatory neurotransmitters such as glutamate encourage the generation of a new action potential in the next neuron.

- Inhibition:

- Inhibitory neurotransmitters like GABA make it less likely that the next neuron will 'fire' and send an action potential.

Every part of this process is tightly controlled and crucial, only happening when it should. Much like a skilled conductor leading an orchestra, each neurotransmitter has a role to play in the grand concert that is our nervous system. The balance between excitation and inhibition is delicate, mirroring the precision required in an intricate ballet. Understanding this elaborate interplay shines a light on how sensations, thoughts, and actions are orchestrated within us, painting a picture of a biological symphony that's both sophisticated and intricate.

Neurons have the crucial job of passing messages through the body, and they do this using electrical signals—think of them as messengers running at lightning speed. When a neuron gets activated by a stimulus—something as simple as the touch of a key or as invigorating as a cold breeze—it undergoes a tiny electrical change. This change is like flipping a switch, causing an electrical pulse, known as an action potential.

The action potential travels swiftly along the neuron's axon, a long, slender fiber that carries the message away from the cell body. The axon acts much like a train track, guiding the electric message on the right path. This message moves fast, sprinting down the axon until it reaches the end of the line, a place called the synaptic terminal.

Here at the terminal, the electrical pulse has to jump across to the next neuron. Since there's a tiny gap between them, the synaptic cleft, the neuron can't just hand off the signal. Instead, it converts the electrical message into a chemical one, using molecules called neurotransmitters. Picture each neurotransmitter as a special code passed in an envelope, carrying the message across the gap.

On the other side, in the neighboring neuron's dendrites, those chemical codes find matching receptors—like a key fits its lock—and the message gets it through. This triggers the electrical signal to start again in this next neuron, cascading the message forward.

This handoff happens over and over, allowing the message to travel through a network of neurons to its final destination, whether that's a muscle to prompt movement or a specific brain region to form a thought. It's a continuous relay race that allows you to experience sensations, think, move, and remember.

Understanding how this all works is like unfolding a map and tracing the pathways that crisscross the landscape of the human body. It's a remarkable journey that happens in the blink of an eye, silently orchestrating every laugh, every step, every heartbeat. In sharing these details, it endeavors not to overwhelm but to enlighten, revealing the beauty and precision of the body's communication system in terms people can relate to and understand.

Let's expand on the depolarization process, an essential phase of the action potential, which is key to how neurons communicate. At rest, a neuron's inner environment is negatively charged compared to the outside. This is due to a careful balance of ions, chiefly sodium (Na^+) and potassium (K^+), on either side of the neuron's membrane, maintained by a structure known as the sodium-potassium pump.

When a neuron is stimulated, specific proteins called sodium ion channels open up first, allowing Na^+ to rush into the cell. This rush flips the charge inside the neuron, making it more positive, which is depolarization. This change in voltage triggers nearby channels to open, creating a wave-like effect as depolarization moves along the axon. It's similar to how dominoes fall one after the other in a chain reaction.

The increase in positive charge inside the neuron isn't permanent. Shortly after the Na^+ influx, potassium channels open, and K^+ moves out of the cell, returning the internal charge to a more negative state—this is known as repolarization. However, immediately following an action potential, the neuron overshoots its negative charge in a phase known as hyperpolarization. The neuron momentarily becomes too negative to fire another action potential.

This is where the sodium-potassium pump earns its keep. It works tirelessly, like a bouncer at a nightclub, pushing Na^+ back out and pulling K^+ back in to re-establish the right levels inside and out. The pump moves

3 Na+ ions out for every 2 K+ ions it moves in, using one molecule of ATP—cellular energy—every cycle. It restores the neuron to its resting state, allowing the neuron to be ready and able to fire an action potential again when required.

This meticulous process of ion exchange and membrane voltage changes is vital because it enables the rapid and precise transmission of signals throughout the nervous system. It's as if each neuron is a relay racer passing the baton flawlessly to the next runner, to ensure the team—your body—responds properly. Understanding depolarization and the sodium-potassium pump shows just how detailed and beautifully orchestrated neuronal communication is, allowing us to grasp the complexity underlying our every move and thought.

Imagine if neurons were the dedicated workers of a bustling city. The dendrites would be the ever-vigilant mail carriers, going door to door collecting letters, the bits of information from the surrounding environment. Once they have the mail, they bring it to the central post office, the neuron's cell body, which acts as the command center. Here, the postal workers sort through the mail, deciding which messages are urgent and need to be dispatched immediately and which can be filed for later.

The axon is akin to the city's highway, a streamlined route where the sorted and packaged messages—now in the form of electrical signals—are dispatched swiftly. Along this route are the axon's myelin sheaths, which act as express lanes, ensuring the information is not just traveling but speedily racing to its destination. At the terminal end of the axon, the messages reach a distribution hub, the synaptic terminal, which faces a narrow river, the synaptic cleft that separates one neuron from the next in the chain.

Here, couriers—neurotransmitters—take the messages and zip across the river in small boats, docking at the receptors, the specialized ports on the receiving neuron. When the message is delivered, it's like a key opening a lock, allowing the next neuron to continue the relay race. Each part of the neuron's complex system mirrors a critical role in a thriving society, ensuring that vital messages keep the city alive, functioning, and reacting to changes with the efficiency and resilience found in nature's most elaborate ecosystems.

Neuroplasticity is a term that describes the brain's remarkable ability to change and adapt. Just as a muscle grows stronger with exercise, the brain can reshape itself when it learns new things or experiences new challenges. It's like reprogramming a computer or using a detour when the usual road is closed – the brain finds new ways to connect its cells, or neurons, to compensate, enhance, or recover abilities.

In more direct terms, when someone repeatedly practices playing the piano or speaks a new language, their brain physically changes; pathways of neurons that are used often become stronger, like forging a new trail through a forest. On the flip side, the less often a pathway is used, the weaker it becomes. This is why habits can be hard to break and why continued practice is key to mastering a skill.

This plastic nature of the brain isn't just for acquiring talents; it's also crucial for recovery. After injuries or strokes, the brain can often reorganize itself, finding alternative paths to get around damaged areas, much like rerouting traffic in a city to bypass roadwork. It demonstrates that the brain is not a static organ but a dynamic, living system, constantly rewriting its own circuitry throughout a person's life.

Neuroplasticity illustrates that learning is not just a mental exercise; it's a physical one, showing the brain's capacity to change and grow, no matter one's age. It's a fascinating process that empowers individuals to understand that with time, effort, and consistent practice, acquiring new skills and habits is always within reach.

Neuroplasticity doesn't just unfold mysteriously; it's rooted in the very tangible dance of molecules and cells in the brain. Let's consider what happens deep within the brain when learning to play a new melody on the piano with daily practice. As fingers press the keys and music fills the air, specific neurotransmitters are hard at work. These chemical messengers include glutamate, which excites the neurons, and GABA, which calms them, preventing excessive activity.

Each time a correct note is played, a circuit of neurons fires, sending electrical signals zipping along the axons, jumping from one neuron to another across tiny gaps called synapses. This repeated firing makes the synapses more efficient, akin to a path in the woods becoming a well-trodden trail. The synapses strengthen - a phenomenon known as 'long-term potentiation,' where neurotransmitter release becomes more effective, and receptors on the receiving neuron enhance their sensitivity.

As the practice continues, the structure of the neurons begins to change. Dendritic spines – tiny protrusions where synapses form on the dendrites – multiply, enlarge, or become sturdier. It's like adding more docks to a seaside town to handle an increased flow of ships. These spines are where the communication between neurons happens, so when there are more spines, and they are more robust, the neural network expands and solidifies, laying down a solid infrastructure for the musical piece being learned.

The changes from practice don't just alter synapses and spines; they can lead to 'neurogenesis,' the birth of new neurons, and trigger shifts in how entire networks of neurons operate together. The brain areas involved in music become more adept, more in sync. Think of it as upgrading a local band to a grand orchestra, with new instruments and more capable musicians contributing to a richer sound.

This cellular-level remodeling is not just about learning music; it's a process that underpins how the brain adapts throughout life – how habits form, how recovery from an injury is possible, and how experiences etch themselves into the very fabric of the brain. Neuroplasticity underscores a fundamental truth about the brain: it is not an inanimate vessel but a living organ that grows and evolves with every challenge and every new skill acquired.

Imagine neurons as the high-speed internet of your body, relaying the click of your mouse or the tap of your finger on a smartphone screen to a vast network, resulting in a video starting to play or a message being sent. These microscopic workers in your brain and throughout your nervous system allow you to feel the warmth of the sun, react to the sizzle of a frying pan, or savor the taste of a perfectly ripe strawberry.

Neurons play a key role in your emotional experiences too, like the rush of joy at hearing a friend's laugh or the surge of adrenaline when you watch a thrilling movie scene. They are the diligent mail carriers of your nervous system, ensuring that every sensory detail, thought, and motion is delivered to the right address without delay.

Even as you sit and read this, neurons are hard at work allowing you to interpret these words, convert them into understanding, and perhaps store them in memory to recall later. They're the unsung heroes behind every deliberate action and automatic reflex – akin to the various apps on your phone working together to keep you connected to the digital world. Neurons are the constant, silent partners in your daily life, translating the world into experiences and keeping you engaged, moment to moment.

To give your understanding of neurons a test drive, think of an activity you can do right now, like clasping your hands together. Notice how effortlessly this motion occurs? Your neurons are communicating from brain to hand in a smooth, coordinated effort, akin to a team of expert rowers gliding across a lake.

Now, try closing your eyes and touching your nose with your index finger. This simple challenge recruits a host of sensory neurons and their motor counterparts, a process similar to a well-choreographed dance. It is a chance to appreciate how your body's neural network helps you navigate space, guiding your movements with precision.

For a mental workout, attempt memorizing a phone number or a short grocery list. Visualize each item or digit. This exercise puts your brain's plasticity into action, paving new neural pathways, similar to laying down a new track for a train to navigate previously unexplored landscapes.

Lastly, consider a sensation such as the texture of your clothing or the ambient sounds around you. Pay attention to the details that these sensory neurons are picking up. It's like tuning into a faint radio station — the more you focus, the clearer the signal becomes.

These activities are not just exercises but an acknowledgment of your brain's dynamic nature. They serve as living examples of neurons at work, showcasing the elegance of the nervous system in orchestrating the symphony of sensory experience, thought, and movement that accompanies everyday life.

In our exploration, we've journeyed through the microscopic landscape of the brain, revealing the neurons' pivotal role as the conductors of the body's symphony. These tireless entities harness electrical signals to communicate, cultivate memories, and enable every sensation and movement. Like intricate wiring in a vast and complex machine, they connect and coordinate, driven by the marvel that is neuroplasticity—showing us that the brain is not just capable of change, but thriving on it, remodeling itself with every new experience.

There is much encouragement to be drawn from the knowledge that every day offers a fresh opportunity to strengthen and forge those neural connections, much in the same way one hones a craft or masters a skill through practice. It's a testament to the resilience and adaptability of our nervous system, a beacon of hope for recovery and growth.

And what a frontier lies ahead! The brain's enigmatic workings are the universe's greatest puzzles, with new secrets waiting to be unlocked. Each discovery within neuroscience offers a glimpse into vast potentials for healing, learning, and even transcending our biological limits. With the tools of science growing ever more precise, and with insights gained from artificial intelligence and computational models, the future is ripe for a deeper understanding of our most mysterious organ.

Take heart in knowing that this journey is far from over. As neuroscience advances, so too does the promise of unraveling the deeper truths about our consciousness, our very nature, and the possibilities for human enhancement. To step into this future is to embark upon the greatest adventure of all—an odyssey not into outer space, but inner space, the last and most intimate frontier.

THE ELECTRIC BRAIN NEURAL COMMUNICATION

Step into the world of 'The Electric Brain,' where the brain's communication system reveals itself to be as enthralling and complex as any language ever spoken or any technology ever devised. At the core of this system are neurons, living threads weaving the tapestry of thoughts, emotions, and actions that make up the human experience. Just as the alphabet forms the basis of language, neurons form the basis of our brain's communication. Each electrical impulse they send is like a letter, each connection a word, each network a complete thought. As this chapter unravels the intricacies of these connections, readers will learn how neurons converse, how they carry the raw data that the mind forms into ideas and memories, and how these tiny sparks of electricity manifest as every sensation felt, thought processed, and decision made. Prepare to be guided through a neurological journey that's both familiar and profound, and discover the silent language that animates our very being. This is the story of how the brain talks, the narrative of neurons' relentless dialogues, and the future of understanding what makes us uniquely human.

An action potential is a fundamental process in neural communication, akin to a wave rushing along the shore before sweeping back out to sea. Imagine a neuron as a quiet beach that suddenly ignites with activity. As a stimulus arrives, channels within the neuron's membrane, much like gates on a dam, open up to let positively charged sodium ions flood in. This rush shifts the electrical landscape of the beach—what was once a tranquil, negatively charged space becomes momentarily positive, creating the wave—the action potential.

This wave travels along the neuron's axon like a surfer barreling towards the beach's end. Once it reaches the synaptic terminal—the end of the line—the action potential prompts the release of neurotransmitters, which bridge the gap to other neurons, continuing the communication process almost like a message being tossed in a bottle across to the next beach.

The action potential's role in neural communication is vital. Each one is a link in an endless chain, a domino that causes the next to fall, enabling the complex cascade of thoughts, movements, and sensations that define human life. This fleeting electrical event is the brain's means of transmitting information rapidly and faithfully across vast networks of neural pathways. It's a brief, yet powerful, spark that underlines every aspect of how we interact with the world around us. Understanding the action potential offers insight into the very pace and pulse of living systems, allowing us to appreciate the silent yet vibrant electrical storm that is perpetually active beneath the calm exterior of the mind.

An action potential is a meticulously orchestrated event that unfolds in several phases within a neuron. It starts at the neuronal resting potential, a state where the inside of the neuron is negatively charged compared to the outside; this is due to a careful balance of ions maintained by the sodium-potassium pump. This pump keeps more sodium ions outside the neuron and more potassium ions inside, much like a gardener might keep certain plants within a greenhouse and others outside.

When a neuron is stimulated, the first major phase, known as depolarization, begins. Sodium channels in the neuron's membrane open, allowing sodium ions to rush inside, shifting the internal charge from negative to positive. Picture this as doors opening during a sale event, customers flooding into a previously calm store.

The peak of this phase is quickly followed by repolarization, where the neuron restores its negative charge. Potassium channels open, letting potassium ions flow out, resetting the electrical balance. Envision the reverse of the situation with shoppers now leaving, the store's normal calm gradually restored.

Sometimes, the neuron overshoots the resting potential in a phase called hyperpolarization, becoming slightly more negative than its initial resting state. It's akin to our metaphorical store being quieter than usual after the rush has passed.

The sodium-potassium pump then resets the ion distribution to the resting state by pumping sodium out and potassium in, establishing the stage to fire another action potential if needed. Think of this as the staff organizing the store, preparing for the next wave of customers.

Once the action potential reaches the synaptic terminal at the end of the axon, synaptic transmission begins. The electrical signal prompts the release of neurotransmitters into the synaptic cleft, the narrow space between neurons. These neurotransmitters travel across to the next neuron, binding to receptors much like a key fits into a lock, prompting the continuation of the signal.

These meticulously synchronized steps—resting potential, depolarization, repolarization, hyperpolarization, and synaptic transmission—are fundamental to neural communication, ensuring that the brain can efficiently process and respond to endless streams of information from the outside world. It's a system that remains one of nature's most impressive engineering feats, underpinning every aspect of thought, behavior, and existence.

Picture the bustling scene of a grand central market, where vendors are lined up with an array of colorful goods. Each stall represents a synaptic terminal of a neuron, ready to dispatch its unique products. These goods are the neurotransmitters, each with a unique flavor and purpose, awaiting their turn to be showcased across the marketplace to the keen customers – the receptors – on the other side of the synaptic cleft, which is the space amid the lively commotion.

When the action potential arrives, it's like the ringing of the market bell, signaling that trading can begin. This signal prompts the vendors – or synaptic vesicles – to release their wares – the neurotransmitters – across the cleft. They surge across like expertly tossed crates landing in the receptive arms of the marketgoers, each catching the package meant for them. The receptors, each a discerning customer, select and bind to the matching neurotransmitter with precision, igniting a response that sends a ripple of excitement through the crowd, much like a command passing swiftly through the bustling market.

This entire lively exchange – from neurotransmitter release to synaptic reception – is the synaptic transmission, a critical process in passing along the baton of information within our neural network. It's a process vibrant with activity, essential for the thoughts, emotions, and movements that compose our every moment. And it's this energetic back-and-forth at the synapse that exemplifies the brain's vibrant conversation, a dialogue that shapes our interaction with the world around us.

Here is the detailed breakdown of the neurotransmitter release and reception process:

- Types of Neurotransmitters and Their Roles

- Excitatory Neurotransmitters:

    - E.g., Glutamate: Typically promote the generation of an action potential in the post-synaptic neuron.

    - Encourage neural activity and communication by increasing the likelihood that the next neuron will fire.

- Inhibitory Neurotransmitters:

    - E.g., Gamma-aminobutyric acid (GABA): Reduce the chances of the post-synaptic neuron generating an action potential.

    - Act as neural suppressants, preventing excessive activity and maintaining balance within the neural circuits.

- Sequence of Molecular Events During Neurotransmitter Release

- Arrival of Action Potential:

    - Triggers the opening of voltage-gated calcium channels at the synaptic terminal.

- Influx of Calcium Ions:

    - Calcium ions enter the synaptic terminal and initiate synaptic vesicles to merge with the synaptic membrane.

- Release into Synaptic Cleft:

    - The vesicle fuses with the membrane, and neurotransmitters spill into the synaptic cleft.

- Diffusion and Binding:

  - Neurotransmitters diffuse across the cleft and bind to their respective receptors on the post-synaptic membrane.

- Types of Receptors

  - Ionotropic Receptors:

    - Act as gatekeepers for ions when the neurotransmitter binds, often resulting in immediate and short-lived effects.

  - Metabotropic Receptors:

    - Influence cell processes through a second messenger system and usually have longer-lasting effects.

- Binding Process and Biological Responses

  - Activation of Receptors:

    - Upon binding, ionotropic receptors open ion channels directly, while metabotropic receptors initiate a cascade of intracellular events.

  - Cellular Response:

    - Leads to either an excitatory or inhibitory effect on the post-synaptic neuron, contributing to the neural network's overall output.

- Modulation and Inhibition of Synaptic Transmission

  - Factors affecting the process:

- Enzymatic degradation or reuptake of neurotransmitters.

- Presence of modulatory substances that can either enhance or inhibit neuronal communication.

- Significance of Variations:

- Essential for the refinement of neural signals, enabling complex functions such as learning, memory, and adaptation to changes.

This complex and nuanced dance of chemicals and signals is the essence of communication within the neural network, with each step crucial for brain function. Modulating this process allows the brain to adapt to new situations and learn from experiences, highlighting the brain's remarkable plasticity and the still-unraveling understanding of its capabilities.

Just as social networks connect people across the globe, neurons link together to form an elaborate network that underlies all brain function. Each neuron is like an individual on a social media platform, capable of sending messages, responding to various inputs, and forming connections that last moments or lifetimes. A neuron sends information in the form of electrical signals down its axon, much like posting a message. When this signal reaches the synaptic terminal, it's as if the message has reached the platform's server, ready to be forwarded across the synaptic cleft—the space that separates one neuron from the next.

The receiving neuron's dendrites, adorned with specialized receptors, are like inboxes, gathering messages and deciding their importance or relevance. These messages trigger responses and become part of the ongoing dialogue within the central nervous system. As in technology networks where devices communicate through protocols, neurons use neurotransmitters and receptors for their exchange.

These interactions at synapses weave a complex, dynamic web of communication that ensures the brain can manage anything from basic body functions to solving quantum physics puzzles. They enlist neurons in distributed processes, collaborating in much the same way cloud computing relies on numerous computers to host services and data. Every learning experience, memory recalled, and emotion felt is a result of these sophisticated, interwoven networks functioning in remarkable harmony.

Understanding this remarkable neural interplay is akin to learning a new language—an endeavor that opens a world of comprehension and connection. It provides the key to deciphering the mechanisms underlying our thoughts and behaviors and the ongoing quest to develop novel approaches to healing and enhancing the brain's function. It's an incessant, vibrant interaction that defines the essence of life, enabling the human experience to be as rich and diverse as it is.

Let's expand on the intricate world of synaptic interactions and network processes. At the crux of this complex dialogue lies the action potential, a surge that ushers information along the neuron. As it races down the axon, voltage-gated ion channels open in sequence, first sodium channels allowing positive ions in and then potassium channels letting them out, restoring the negative charge. It's a carefully timed dance, controlled and precise, balancing inputs and outputs to propagate a signal.

Upon reaching the synaptic terminal, a remarkable transformation takes place. Here, synaptic vesicles loaded with neurotransmitters meld with the terminal membrane, releasing their cargo into the synaptic cleft. Picture a lock sliding open, releasing a flood of water through a canal gate. This neurotransmitter release is critical; it is the signal that hops the fence between neurons.

Various receptors await these signals, each specialized to interpret specific neurotransmitter types, like radio stations tuning into different frequencies. Excitatory receptors amplify signals, propelling the action forward, while inhibitory receptors dull the response, applying the brakes to neuronal activity. They decide if the receiving neuron will fire its own action potential and continue the conversation.

Summation is how a neuron integrates these diverse incoming signals. Consider how a team assesses multiple, sometimes conflicting, strategies before making a company decision. Both temporal summation—multiple signals arriving in quick succession—and spatial summation—simultaneous signals from multiple locations—can culminate in a decision for the neuron to fire, consolidating information into an orchestrated response.

Delving into the nuances of synaptic plasticity uncovers the basis for learning and memory. Long-term potentiation fortifies the signal between two neurons when used frequently, like strengthening a friendship with repeated positive interactions. Conversely, long-term depression weakens the connections due to disuse or negative association, akin to a path overgrown with weeds when not traveled. These processes are foundational for the adaptability and learning capability of the brain.

In sum, these interactions build not just responses, but experiences and memories, making them cornerstones of neural communication and brain function. Each plays a role in crafting the dynamic story of our interactions and responses to the world. It's a vibrant conversation written not in words, but in the electric and chemical language of the body.

Imagine you're in a bustling kitchen where a symphony of chefs is whipping up a feast. Each chef represents a neuron in the brain's neural network, and their recipes are the electrical and chemical signals they send out. Just as the timing of adding ingredients to a pan is crucial for the perfect dish, so too is the timing of these signals for a well-functioning neural network.

The electrical signals, much like a chef's quickfire commands, shoot down the neuron's axon to reach other members of the culinary team at the synapses. The moment these electrical signals reach their destination, they switch from shouting orders to passing notes – these are the chemical signals, or neurotransmitters, which float across the gap between neurons like an aroma wafting between kitchens.

Each neuron waits for these scent-filled notes, choosing the right moment to add their spice to the pot – this is how chemical signals influence the next neuron's action potential. Just as too much salt can ruin a soup, too much neurotransmitter can overstimulate a neuron. Likewise, just as a dash of sugar can balance a dish, the right amount of neurotransmitter can create the perfect neural response.

And so, in this bustling kitchen of the brain, electrical and chemical signals must be precisely timed and measured, ensuring each dish – each thought, sensation, or movement – turns out just right. It's a delicate balance, an intricate dance of signals, demonstrating not just the elegance of the brain's inner workings, but its astounding capability to produce the rich tapestry of the human experience.

As neuroscience peels back the layers of the brain's mysteries, each discovery acts as a stepping stone towards groundbreaking technologies and transformative medical treatments. Take, for example, deep brain stimulation, a technique akin to fine-tuning a radio to the perfect frequency, which can alleviate symptoms for Parkinson's patients. This sophisticated method offers a glimpse into how refining neural communication could restore movement and improve quality of life.

Advancements in neuroimaging, the brain's version of satellite photography, allow doctors to navigate neural landscapes with precision, spotting the early signs of diseases like Alzheimer's and potentially halting them in their tracks. Similarly, brain-computer interfaces — akin to tapping directly into the control panel — promise to grant mobility to those with spinal injuries by translating thought into action.

On a more everyday level, these neuroscientific findings influence the development of apps designed to bolster mental health, enhance learning, or even fine-tune sleep patterns. It's like having a personal cognitive coach or sleep conductor in one's pocket, a tool that can turn the cacophony of an overstimulated mind into a harmonious melody of calm and concentration.

These innovations embody the potential of applied neuroscience. Like a bridge linking two shores, the gap between understanding the brain's functions and improving human lives becomes narrower with each step forward in research. Future neurotechnologies may not just mend but could also augment human ability, inviting people to experience life with greater vibrancy and resilience. This progress is not just a victory for science; it's a beacon for anyone touched by the far-reaching hand of neurological challenges, providing hope that life can not only go on but also flourish.

Alan Turing's work laid the foundations for modern computing, much like an architect preparing the blueprint for what would become a vast city of computers. His concept of a universal machine—later known as the Turing machine—presented the idea that a single system could emulate any other, just as complex neural networks in the brain can process diverse types of information.

Ramon y Cajal, often referred to as the father of modern neuroscience, provided the first detailed descriptions of neurons. He proposed that these cells were separate entities, not fused into a continuous web—a concept called the neuron doctrine. This idea revolutionized the understanding of the brain, revealing it as a tapestry of discrete but interconnected units, similar to Turing's discrete machines that form a comprehensive system.

Both Turing and Cajal's insights were pivotal in transforming their respective domains. They helped unravel the inner workings of systems that at first glance appeared impenetrable. Turing's vision gave rise to the computers and algorithms that we rely on daily, while Cajal's discoveries opened the door to deciphering the brain's communication patterns, leading to the development of treatments for neurological disorders and enhancing our understanding of how we think, learn, and remember.

The parallels between their work underline a fundamental truth: complex networks, whether artificial or biological, are built upon the interaction of simple units—be they silicon-based transistors or carbon-based neurons. Each unit's connection and the flow of information within the network define both a computer's capability and the brain's cognitive prowess. Looking at their contributions, it's clear that both Turing and Cajal did not just map out uncharted territories, but they also provided the key to navigating these landscapes, thus enabling future innovators to build upon their legacies.

Let's expand on the pioneering work of Alan Turing and the detailed mechanics of the Turing machine. A Turing machine is a hypothetical computing device, conceptualized in the simplest form with a tape divided into cells, each capable of holding a single symbol. At any point, the machine has a 'head' that can read and write symbols one at a time, moving the tape left or right as it follows a set of predetermined rules or 'states'. This rudimentary construct can simulate any computation process—a testament to Turing's brilliance in encapsulating the essence of algorithmic execution.

Now, delve into the neuron doctrine by Santiago Ramón y Cajal. This principle, challenging existing beliefs, posited that neurons are discrete entities that communicate via points of contact—now known as synapses—rather than a continuous network. This shifted neuroscience paradigms, framing the brain as an intricate mosaic of individual yet interconnected cells, setting in motion a wave of research that has since unearthed the vast complexities of neural connectivity and communication.

When comparing these conceptual frameworks, one can see structural similarities and distinctions between artificial and biological networks. Both consist of discrete units (transistors/neurons) that process and transmit information. Yet, while Turing machines are sequential and binary in their operations, neural networks perform parallel, analog computations, showcasing an astounding plasticity.

From Turing's abstract machines rose modern computers, algorithms, and even the field of artificial intelligence—machines emulating cognitive processes, once thought the sole dominion of living organisms. Cajal's work laid the groundwork for advances in understanding neurodegenerative diseases and the development of neuroimaging technologies, peering deeper into the brain's workings than ever before.

By exploring these foundational theories, one gains a richer perspective on both the history and the trajectory of computational and neurological science. Alan Turing and Santiago Ramón y Cajal did not just establish key concepts; they engineered the lenses through which humanity perceives, and invariably alters, the realms of technology and medicine. Their legacies promote not mere comprehension but imagination, inspiring generations to continue unraveling the enigmas of the mind and the potential of its synthetic counterpart.

Wrapping up this chapter, one steps back to marvel at the brain's intricate network, an orchestra of cells that composes the symphony of human thought, emotion, and movement. This journey through the landscape of neurons, synapses, and neurotransmitters has revealed a masterpiece of biology, one that orchestrates the complex ballet of neural communication. Each electrical impulse and chemical message is a note in the melody of the mind, contributing to the grand narrative of the human experience.

Taken together, these processes are not simply mechanisms; they encapsulate the essence of being—how one savors a meal, feels a breeze, remembers a loved one, or dreams up an invention. It's in this dance of impulses and reactions that one finds the roots of creativity, learning, suffering, and joy. From the steady drum of resting potentials to the crescendo of action potentials, and the delicate whisper of neurotransmitter exchange, the brain communicates within itself and with the body in a language more intricate than any ever spoken.

Regardless of whether you are a novice to the fold of neuroscience or carry the torch further along the path of discovery, this chapter invites you to contemplate the quiet wonder of the brain's communications. It seeks not only to inform but to inspire a sense of awe and respect for the profound complexities tucked away within the three pounds of matter in the skull. With ever-advancing research and technology, the journey of understanding how the brain communicates is boundless, full of potential to unlock further secrets of our nature, and, ultimately, to enhance the tapestry of life itself.

BRAIN ARCHITECTURE ORGANIZATION AND FUNCTION

Step inside the magnificent structure of the human brain, a marvel of nature as intricate as it is enigmatic. At first glance, one might liken this organ to a beautifully complex circuit board, pulsating with electricity, alive with signals zipping back and forth. Yet, this is no mere tangle of wires; it's an elegant network, with each neuron, axon, and synapse meticulously arranged to facilitate the myriad functions that shape who we are, how we act, and why we feel.

Like a masterful conductor, the brain orchestrates this symphony of neural activity, ensuring that every laugh, every heartbeat, every memory is a testament to its design. Within this organ lies the power to craft poetry, to solve puzzles that have mystified generations, to hold the image of a loved one long after they have departed. The brain's architecture isn't just about the neurons themselves; it's about the connections they make, the patterns they weave, and the stories they tell within the vast theater of the mind.

As one digs deeper into this topic, it becomes clear that the brain's influence on human endeavor stretches far and wide. It is the silent force behind every invention, the unseen spark of every idea. To understand the brain's structure is to gain insight into the very fabric of human existence, unlocking secrets that reside in every fold and crevice of this extraordinary organ. This chapter serves as a map, guiding those who seek to comprehend the potent yet delicate interplay that forges our consciousness and carves out our place in the universe.

Picture the brain as a sprawling metropolis, each region a distinct district buzzing with its own unique activity. The cerebrum, with its rolling hills of cerebral cortex, is the downtown core—the center of business and culture—where executive decisions are made, and creative ideas bubble like coffee in a busy café. Here lie the offices where strategies are formed, much like the frontal lobes process complex thought and planning.

Head over to the cerebellum, the recreational area of our city, with lush parks where people refine their dance steps or practice tai chi. It's akin to how the cerebellum fine-tunes our movements, maintaining grace and coordination. Then, there's the brainstem, resembling the bustling infrastructure beneath the city, running unseen yet vital—managing the flow of traffic, water, and electricity to keep the metropolis alive.

Each part of the brain, just like every zone in a city, plays its role. The limbic system, home to our emotions and memories, could be the cultural heart of the city, with its art galleries and theaters capturing and expressing the raw human experience. And not to forget the suburbs—the basal ganglia, responsible for habitual actions and procedural learning, much like how suburban life involves routines and familiar patterns.

Through this urban lens, the brain's regions become not just relatable but vividly alive, illustrating not merely the what and how of brain function, but painting a picture of why these activities are central to the essence of who we are. It's a brainy city where every citizen, idea, and moment has its place, contributing to the grand narrative of life.

Here is the detailed breakdown on the cerebral cortex's components and their functions:

- Lobes of the Cerebral Cortex

  - Frontal Lobe:

    - Responsible for high-level cognition, including reasoning, planning, problem-solving, and decision-making.

    - Houses areas vital for voluntary motor activities, like the precentral gyrus, or 'primary motor cortex.'

    - Contains the prefrontal cortex, which is crucial for personality expression and moderating social behavior.

  - Parietal Lobe:

    - Integrates sensory information from various modalities, primarily determining spatial sense and navigation (proprioception).

- Plays a role in the management of objects we perceive and their physical properties.

- Houses the somatosensory cortex, processing tactile information from the body.

- Occipital Lobe:

- Main center for visual processing; involves the identification and categorization of objects and the perception of motion and color.

- Contains the primary visual cortex, where visual stimuli begin to be interpreted.

- Engages in higher-level visual processing, like comprehension of visual information.

- Temporal Lobe:

- Involved in auditory processing, including the understanding of spoken language.

- Contains the hippocampus, instrumental in forming new memories.

- Helps with organizing and retrieving memory, appreciating music, and discerning faces.

- Executive Decision-Making

- Frontal Lobe Circuits:

- Dorsolateral prefrontal circuit plays a key role in managing executive functions like working memory and attentional control.

- Orbitofrontal circuit helps in evaluating risks and rewards, essential in decision-making.

- Anterior cingulate circuit is involved in motivation and error detection during complex tasks.

- Sensory Information Integration

- Parietal Lobe Functionality:

- Merges visual, auditory, and tactile signals to provide a comprehensive understanding of the environment.

- Facilitates complex tasks such as writing and drawing by interpreting spatial orientation and body position.

- Visual Processing

- Occipital Lobe Functionality:

- Primary visual cortex receives input from the eyes and begins the initial stages of visual interpretation.

- Dorsal and ventral streams further process the 'what' and 'where' aspects of objects seen, enabling recognition and spatial awareness.

- Memory and Language

- Temporal Lobe Contributions:

- Wernicke's area, crucial for language comprehension, permits fluent speech and understanding of spoken words.

- The hippocampus, laying down new episodic memories and facilitating spatial memory which enables navigation.

Each component of the cerebral cortex is like a gear in the complex machinery of the mind, working together to produce the seamless experience of human thought, sensation, and action. Understanding these parts with precision is like reading a map before a journey—vital for navigating the vast landscape of brain function and appreciating the subtle and coordinated processes that empower every aspect of our daily lives.

In the vast expanse of the brain, each neuron operates as a messenger, carrying vital information across the nervous system. Picture a neuron as a high-tech communication device, intricately designed to receive, process, and transmit data. It gathers signals through its dendrites, tree-like extensions that act like antennas picking up broadcasts. These signals then converge in the neuron's cell body, the central hub that decides what message needs to be passed on.

When a decision is made, the neuron fires an action potential – this is the language of the nervous system, a surge of electrical energy that rapidly travels down the length of the axon, a cable-like projection. The axon is wrapped in an insulating layer called myelin, which, like the casing on electrical wires, ensures that the signal zips along without losing strength.

At the axon's end, the action potential triggers the release of neurotransmitters, substances that act as biological couriers. These molecules cross the synaptic gap – the small space separating neurons – to reach the neighboring neuron's receptors, much like a baton handed off in a relay race. The receiving neuron interprets this chemical message, and the cycle begins anew.

Through this remarkable relay of electrical and chemical signals, the brain communicates within itself and with the body, forming a network that is as complex as it is efficient, allowing us to think, feel, move, and experience the world in all of its splendor. Getting a firm grip on how a neuron works is akin to understanding the first note in a vast symphony – it's the beginning of a journey into the incredible orchestration that is human consciousness.

Neurotransmitters are brain chemicals that communicate information throughout our brain and body. They relay signals between nerve cells, called neurons. To understand their action and influence on neural processing, a breakdown into categories is helpful.

There are excitatory neurotransmitters, like glutamate, which promote the creation of an electrical signal called an action potential in the post-synaptic neuron. In simple terms, they are like the brain's green light, initiating activity. Inhibitory neurotransmitters, such as GABA, do the opposite; they block signals and help maintain equilibrium in your brain's activity—consider them the brain's red light. Modulatory neurotransmitters, such as dopamine and serotonin, can either amplify or diminish the signals, re-tuning the neural pathways for optimal performance.

Now, here's what happens at a microscopic level during neurotransmitter release and receptor binding: When an electric signal reaches the end of a neuron, it prompts small packets containing neurotransmitters to fuse with the neuron's membrane, pouring their contents into the space between neurons, known as the synaptic gap. These neurotransmitters then bind to receptors on the surface of the next neuron, which can either start a new action potential, if they are excitatory, or prevent one, if they are inhibitory.

Synaptic plasticity is the ability of synapses, the connection points, to strengthen or weaken over time, in response to increases or decreases in their activity. Synaptic plasticity is considered one of the primary cellular mechanisms that underlies learning and memory. Each time you learn something new, your brain slightly shifts the strength of the connections at work, making it easier to activate that particular pathway the next time.

Imbalances in these neurotransmitters can lead to various mental health conditions. For instance, a deficit in serotonin is associated with depression, while dopamine dysregulation is linked to schizophrenia and bipolar disorder. If you imagine the neurotransmitters as messengers carrying crucial information, when the messages are too few or too many, or arrive at the wrong time, it can disrupt the entire communication system.

Imagine the brain's network of neurons much like the vast roadway system of a bustling city. Neurons are the roads and highways, each axon a lane zipping information to its next destination. The electrical signals are cars, traveling at high speed to deliver messages that dictate everything from the tap of a toe to the spark of a thought. At each junction, or synapse, these signals can switch routes, like taking an off-ramp to another freeway, ensuring they reach precisely the right neighborhood of the brain.

Just like street signs and traffic signals guide drivers through the city, various proteins and chemicals in the brain direct neural signals, ensuring they flow smoothly and accurately. This prevents traffic jams or accidents in information conveyance within the neuro-metropolis. And when there's heavy rain or a roadblock, akin to injury or disease, traffic might slow down or take detours—a reflection of the brain's remarkable ability to adapt and reroute when faced with challenges.

The importance of these neural pathways goes beyond biological functions—they are the roads on which our consciousness navigates, the routes that thoughts, emotions, and memories travel to become actions and experiences. The elegance lies not only in the complexity of this vast network but also in its efficiency and adaptability, showcasing the extraordinary nature of the brain's infrastructure. Each message's journey along these neuron pathways is as crucial as the destination it reaches, shaping who we are and how we interact with the world.

Let's take a deeper look at the dynamics of synaptic transmission, revealing the nuances of this critical communication process within neural networks. Neurons converse through neurotransmitters, each carrying its unique message, much like vehicles on city roads have different purposes—the ambulance rushes with urgency, the cargo truck delivers goods, and the bus transports passengers.

In the brain, excitatory neurotransmitters, such as glutamate, are the urgent messages that actively tell receiving neurons to 'go' or activate, akin to pushing the gas pedal. Inhibitory neurotransmitters, like GABA, send the 'stop' signal, just as red lights control traffic flow. The neuromodulators, such as dopamine and serotonin, serve as regulation adjustments, sometimes ramping up activity or dialing it down, akin to adjusting traffic patterns during peak and off-peak hours.

The process of synaptic transmission begins when an electrical impulse, or action potential, moves down an axon to the synaptic terminal. There, synaptic vesicles, which hold neurotransmitters, dock at the terminal's edge like buses at a station. Triggered by the action potential, these vesicles release their neurotransmitter contents into the synaptic gap—the road across which the 'vehicles' must travel. The neurotransmitters bind to specific receptor sites on the neighboring neuron, functioning as the keys to ignition, sparking activity in the next neuron, or, alternatively, acting as the brakes.

Synaptic plasticity is like the city updating and improving its roads. With frequent use, neural pathways become more robust, mirroring the familiar roads that reinforce with repeated travel—a physical reflection of learning and memory. Conversely, less-traveled paths can weaken or get rerouted, a testament to the brain's flexibility and capacity for change.

Imbalances in neurotransmitters can disrupt the entire neural communication network, akin to a major city's traffic snarl-up from accidents or roadworks. A surplus of excitatory neurotransmitters might result in a frenzied burst of activity, like a traffic jam at rush hour. Conversely, a deficiency may lead to lethargy and disengagement, resembling a city at a standstill during a blackout. These imbalances can manifest as mood swings, anxiety, learning difficulties, or other neurological conditions, illustrating how delicate the balance of this neural traffic must be to maintain optimal functionality.

This intricate procession of chemical messengers along synaptic highways is fundamental to understanding not just brain function but our very experiences and behaviors. By examining each component of synaptic transmission in this way, we uncover a clearer, more complete picture of how the brain relays messages that form the fabric of thoughts, emotions, and ultimately, the human experience.

Think of the brain's cognitive territories as various hubs of a bustling city, each dedicated to an array of vital operations. The frontal lobe, often associated with complex planning and decision-making, is like the city hall—here, the executive decisions are taken, the big plans are drawn out, and the future of the city is shaped. It's where impulse control and judgment call the shots, much like a seasoned mayor overseeing the functions of a metropolis.

Venture into the parietal lobe, akin to the city's central post office, it's a hub where sensory information is processed and dispatched, allowing for the navigation and manipulation of objects. Whether it's reading a map or sending out parcels to the correct address, the parietal lobe ensures that touch, spatial orientation, and coordination are processed and interpreted efficiently.

The occipital lobe, the city's cinema, translates light and shadow into stories and scenes, making sense of the visual signals that are constantly streaming in. Like an experienced film editor, the occipital lobe cuts and crafts the raw footage from the eyes into the coherent movies we perceive as our visual experience.

Then there's the temporal lobe, reminiscent of a concert hall, resonating with sounds and harmonies. It houses our hippocampus—like an archival library of sounds and sights—that keeps memories alive across time, encapsulating the essence of our lives like a timeless melody.

Each region of the brain, similar to these civic centers, carries out its unique function, essential to the workings of the mind's metropolis. Much like city infrastructure, these cognitive areas need to work in harmony, ensuring that every thought, decision, and sensation contributes to the grand tapestry of human consciousness. Each area has its role, each process its purpose, coming together to form the vibrant, pulsating rhythm of our daily lives.

Here is the breakdown of the frontal lobe's components and their contributions to complex planning and decision-making:

- Key Areas within the Frontal Lobe

  - Prefrontal Cortex:

    - Acts as the executive of the brain, processing complex cognitive behavior, personality expression, decisions, and moderating social conduct.

  - Broca's Area:

    - Specializes in the production of language, orchestrating the speech muscles for fluid communication.

  - Motor Cortex:

    - Coordinates and executes voluntary movements, sending out instructions to the body similar to a dispatcher.

  - Role of the Dorsolateral Prefrontal Cortex

    - Manages executive functions such as working memory, helping juggle different pieces of information while problem-solving.

- Oversees attention to tasks, much like a conductor ensures each section of the orchestra plays at the right moment.

- Decision-Making Sequence

- Information is received from the sensory areas and relayed to the prefrontal cortex, where it's weighed against memories and potential outcomes, similar to a council assessing a new urban plan.

- Decisions are formulated based on this multitude of data, akin to a mayor picking the best course of action for the city's welfare.

- Function of the Primary Motor Cortex

- Translates decisions into action, transmitting meticulous movement orders to the muscles, as precise as a coordinator planning the city parade route.

- Significance of Broca's Area

- Facilitates the physical aspects of speech, crafting sentences as a speech writer would craft a compelling address.

- Integral to decision-making during conversations, enabling quick and appropriate verbal responses, like a press secretary representing a city.

Analogies to City Governance

- Imagine the prefrontal cortex as the governor's office, where broad overarching decisions about the city are made.

- Broca's area serves as the communications department, developing the messages that the city broadcasts to the public.

- The motor cortex represents the city's public works, laying down the plans and then putting them into motion efficiently.

Each of these regions plays a distinct but interconnected role in how the brain processes complex thought and translates it into action. By understanding these functions and their interrelations, one gains deeper insight into the brain's remarkable capability for nuanced thought and the execution of sophisticated tasks. Like a city's infrastructure that supports the flow of people and information, each area of the frontal lobe supports the flow of neural data pivotal to human experience.

Consider how a well-planned event comes to life—the brain's output of behaviors can be seen in much the same light. Imagine the mind is like a seasoned event planner, meticulously orchestrating every detail from the guest list to the evening's schedule. The thoughtful planning represents cognitive processes, where every aspect is pondered, from the theme of the event to the timing of each speech. The moment the doors open and the guests begin to arrive, these plans transition into actions, just like thoughts and decisions manifest as behaviors.

In professional settings, it's like a project manager turning blueprints into a standing structure. The brain assembles the building blocks of thoughts, goals, and skills, laying down the foundation. Then, with the precision of a craftsman, translates those plans into the tangible outputs: the decisions we make, the words we speak, and the tasks we complete.

In social spheres, it's akin to the dynamics within a group of friends deciding on a collective experience. Someone suggests a movie night, and just like that, neurons firing off in the depths of the intellect set off a chain of communications and actions that culminate in laughter and popcorn shared on a comfortable sofa.

Every behavior the brain produces, from the subtle to the grandiose, is a result of intricate backstage work that happens in the neural network, akin to the unseen efforts that ensure a performance or a conference runs without a hitch. This parallel offers a glimpse into the beauty and complexity of how our inner experiences are expressed outwardly, shaping our interactions with the world and crafting the stories of our lives.

Let's take a deeper look at the intricate journey from thought to action within the neural tapestry of the brain. This process begins with an initial spark—a thought, taking form in the prefrontal cortex, where decisions are weighed and intentions are formed. Picture a neuron here as the architect drafting the blueprint; it's where the concept is conceived.

   - Neural Pathway to Action

   - The prefrontal neuron relays a burst of electrical impulse to associative neurons, like a designer passing the blueprint to the engineers.

   - Associative neurons, responsible for processing complex input, refine the plan integrating sensory information, memories, and emotions.

   - Instructions are sent to the motor cortex, which acts like the construction foreman, translating plans into clear movement orders.

   - Electrical Impulses to Movements

   - Motor neurons receive the brain's electrical commands and convert them into chemical messages.

   - These chemical messages, neurotransmitters, jump across synaptic gaps to muscle fibers, signaling them to contract much like workers laying down the actual framework of a building.

   - Role of Neurotransmitters

   - Neurotransmitters are the messengers, coordinating neuron-to-neuron communication, akin to a city's utility systems delivering services precisely where needed.

   - Excitatory neurotransmitters move the action along; inhibitory ones pause it, providing a check-and-balance like traffic lights ensuring smooth flow.

   - Feedback Mechanisms in Behavior

   - Sensory feedback is constantly sent back to the brain, like a site manager relaying progress reports.

- The brain adjusts motor output accordingly, in real-time corrections and adaptions resembling an event planner's on-the-fly tweaks for the perfect timing and flow.

This microcosm of activity, from a single neuron sparking an idea to a muscle fiber contracting to create movement, paints a clear picture of the brain's incredible ability to turn abstract thoughts into concrete actions. It's the symphony of countless such pathways that enable us to interact fluidly and adeptly with our environment, bearing in mind that this simplified explanation merely brushes the surface of the brain's complex functionality. Each neuron's contribution is vital, each process essential, for the seamless execution of tasks that we navigate daily in our personal and professional lives.

Through the annals of scientific discovery, some figures stand out for their contributions to our grasp of the brain's mysteries. One cannot speak of the brain's inner workings without tipping the hat to Santiago Ramón y Cajal, the trailblazing Spaniard whose intricate drawings of neurons shifted the paradigm of brain structure, cementing the idea that the brain is not a continuous network but a constellation of individual cells.

Then there's Wilder Penfield, a pioneer of brain surgery, who, through meticulous electrical stimulation during surgeries, mapped the brain's significant regions. This is akin to charting a new continent, with Penfield marking down the rivers of sensory function and the mountains of motor control on an ever-developing map.

Roger Sperry's split-brain research is yet another milestone, unraveling the diverse functionalities of the brain's hemispheres like a maestro understanding the unique sounds of the instruments in an orchestra. He showed that each hemisphere plays a different tune in the symphony of cognition, contributing uniquely to our perception and behavior.

These explorers, each with their lantern of curiosity, ventured into the brain's dark recesses and emerged with findings that lit the way forward. They not only expanded the frontiers of knowledge but also opened doors to treatments that have restored function and hope to many. Even today, their torches light the paths of new explorers delving into the brain's depths, proving that understanding the structure and function of the most complex organ is a journey far from over, but one where each step has the potential to illuminate untold mysteries of the human experience.

The functional architecture of the brain can be seen as the original blueprint for many technological wonders we rely on today. Just think of the computer's central processing unit (CPU), the brainchild inspired by the brain's own intricate network. The cerebral cortex acts like a sophisticated CPU, processing vast amounts of data, making decisions, and running the show. Both are powerhouses of operation, conducting complex tasks and regulating a myriad of activities simultaneously.

Consider the similarities between the brain's neural networks and artificial neural networks in machine learning. Just as our neurons connect through synaptic links that strengthen with learning, artificial neural networks use algorithms to adjust digital connections, learning from patterns and improving over time. These artificial networks mirror learning and adaptation—replicating our own abilities to recognize faces or understand language.

Even the internet, with its web of connections allowing global communication, reflects how the brain's vastly interconnected neurons facilitate internal chatter across various regions. It's as if each website, email, or social media platform has its counterpart in thoughts, memories, or sensory experiences crisscrossing within the brain's elaborate labyrinth.

The brain's knack for parallel processing is mirrored in modern supercomputers that handle multiple tasks at once, a testament to the brain's efficiency and the inspiration it continues to provide. These innovations, fueled by the brain's functional architecture, are not mere replications; they are testaments of ingenuity, recognizing patterns in biology and using them as a leaping point for technological advancement. Through bridging these worlds, the maze of neurons in our heads doesn't just help us navigate the complexities of life; it also sparks the creation of tools that extend the possibilities of human experience.

Let's take a deeper look at the cerebral cortex and draw parallels with the central processing unit of a computer to illuminate their respective roles and functionalities:

- Primary Areas of the Cerebral Cortex:

  - The prefrontal cortex serves as the decision-making hub, sorting through various options and outcomes, akin to an executive in a company making strategic choices.

- Motor areas of the cortex are responsible for planning and executing movement; consider these as the hands that carry out the executive's directives, putting plans into action.

- Parallel Functions within a CPU:

- The control unit of a CPU directs operations like the prefrontal cortex oversees decision-making processes, managing the flow and execution of tasks.

- An arithmetic/logic unit (ALU) is responsible for carrying out mathematical operations and logical decisions, mirroring the precision of the motor cortex in action.

- Comparison of Neural Transmission and Data Pathways:

- Neurons transmit signals to facilitate thoughts and behaviors, comparable to how a CPU receives and processes data to perform complex tasks.

- Both systems rely on well-established pathways to ensure the accurate and efficient relay of information, whether it be electrochemical signals or binary data.

- Brain Plasticity and Machine Learning Algorithms:

- The brain's ability to adapt and rewire itself, known as neuroplasticity, is similar to how machine learning algorithms optimize their performance over time through experience and accumulated data.

- Both systems effectively 'learn from experience,' enhancing their problem-solving abilities and refining the processes they undertake.

- Parallel Processing Capabilities:

- Just as the brain can manage multiple tasks concurrently through parallel processing, managing sensory input, thoughts, and movements all at once, supercomputers execute numerous operations simultaneously, thanks to multi-core processors.

- This ability to multitask efficiently is critical to the functionality of both biological and artificial systems, demonstrating an elegant synergy between the inherent capabilities of the brain and the engineered prowess of technological advancements.

The cerebral cortex's influence on computing technology is a testament to human ingenuity, with each discovery in neuroscience potentially ushering in a new era of computational power and ability. Understanding each component's role and how they work together provides an appreciation of the brain's complexity and the technological achievements it has inspired. This intricate knowledge assists not only in comprehending our own biological makeup but also in forging the path to future innovations that can enhance our lives.

In the odyssey of understanding the brain, one finds that its sophisticated organization is elegantly reflected in the world's intricate technologies. From skyscrapers that rise like the layers of the cerebral cortex to social networks that intertwine like neural pathways, each creation is a testament to human innovation, inspired by the biological marvel behind one's very eyes.

Consider this finale not an end but an invitation to peer deeper into the brain's labyrinth and recognize its echoes in our daily lives and works. Modern marvels don't just imitate the brain's intricacies; they reveal a profound connection between natural and synthetic realms, underscoring the potential for even more revolutionary breakthroughs.

As you close this chapter, let the marvels of the mind stir a curiosity that reaches beyond the confines of the known, kindling an explorative spirit. For in the dance of neurons lies the choreography for future wonders, and in the patterns of thought, the blueprints for tomorrow's innovations.

This journey into the brain's essence and its influence on human ingenuity offers more than just facts—it unveils a story of transcendent connections, encouraging you to engage with the world not just as participants, but as pioneers at the frontier of the great unknown.

SENSORY SYSTEMS AND PERCEPTION

Get ready to understand the astonishing panorama of the human sensory systems, the intricate network that orchestrates one's interface with the world. Each moment, the senses gather streams of information, transforming sights, sounds, scents, tastes, and textures into the rich tapestry of what one perceives as reality. This journey is about to unfurl the miraculous processes behind the everyday experiences often taken for granted.

In these pages lies a story not of mere biological function but of perception, the alchemy by which the brain reimagines external stimuli into subjective experiences. The aim here is to strip away the layers of this remarkable process, exposing the precise mechanics of how a rose is not only seen but also smelled and how a melody is heard and felt, resonating within.

With clear language and without leaning on dense jargon, the aim is to rend the veil of complexity that shrouds the senses. The narrative unfolds like a map, guiding one across the terrain of neurobiology, using vibrant examples and relatable analogies to illuminate each concept. Just as a painter blends colors to bring a scene to life, this exposition blends scientific fact with eloquent explanation, bringing the unseen concert of sensory perception to light.

So this journey begins, with a promise to equip the reader with not just knowledge but also wonder, making transparent how each received whisper of sensation contributes to the symphony of perception. This is more than mere translation from impulse to experience; it's an invitation to understand the profound artistry behind the senses, an artistry that colors every moment of existence.

Imagine the body as a grand mansion, with the five senses as its windows to the outside world. Sight is the picture window, offering a panoramic view of the world in color and motion, vital for guiding movement and recognizing faces. Hearing is likened to an intercom, bringing in a symphony of sounds from thunderous booms to whispers, key for communication and alerting to changes in the environment.

Touch is the home's tactile interface, replete with sensations of pressure and temperature, essential for feeling the warmth of a hug or the pain of a burn, woven into the very fabric of physical interaction. Taste and smell are the gourmet kitchen's sources of flavor and scent, creating a rich palette for savoring meals and detecting potential dangers, like the foul odor of spoiled food.

Each sense feeds streams of data to the brain, the mansion's eloquent host, which interprets this raw information into what one perceives as reality. Together, these senses function as a life-enriching system, as crucial for survival—steering clear of danger, finding food and shelter—as they are for the joys of living, like appreciating music or savoring a delicious meal. The sum of their work shapes not only perception of the environment but also the foundation of memories and experiences, crafting the narrative of life itself.

Let's unfold the remarkable process of how the brain pieces together the signals from the five senses to form a unified, multidimensional perception of reality. Begin with the detection of a sensory signal, much like a welcome chime that announces a guest's entry into the grand mansion. Each sense has specialized receptors that act as the doorkeepers, picking up specific types of sensory information from the environment.

When a guest—our sensory signal—enters, they're met by these receptors which then translate their unique characteristics, such as light for sight or sound for hearing, into neural signals. These signals are like the pieces of gossip that flutter through the mansion, carrying the news from room to room.

These whispers of information reach a crucial juncture at the thalamus, the mansion's central switchboard, which directs the neural signals to various rooms—the different regions of the cerebral cortex. Here, in the mansion's master control room, the cerebral cortex, the brain takes in the flood of incoming data, weaving together sensations to form perceptions.

Imagine the kitchen where culinary magic happens: various ingredients are combined to create one harmonious dish. Similarly, the brain combines signals such as color, shape, and depth from the eyes with texture, temperature from touch, orchestrating them into the recognized form of, say, an apple.

Take the example of smelling freshly baked bread. The olfactory receptors send the scent message straight to the brain, bypassing the thalamus. The aroma might whisk you down memory lane to a cherished moment in your grandmother's kitchen. That's because the brain links this smell with past experiences and their accompanying emotions, all stored in the brain's memory vault.

Through each step, from reception to integration, the brain's mastery in generating perception is apparent. It's a complex dance of signals and responses, beautifully choreographed to construct the reality you experience—a reality rich with color, sound, texture, taste, and aroma, synthesized into a single vibrant scene setting the stage for your memories and emotional responses. Understanding these steps in the brain's grand ballet of perception brings one closer to unraveling the grand narrative of human experience.

Embark on a sensory journey akin to savoring a masterfully composed dish at a fine dining restaurant. Each flavor you encounter is like a sensation, an isolated note that dances on your taste buds, much as a single instrument plays its tune in a song. The tang of citrus, the warmth of spice, the richness of butter - each taste arrives on your palate distinct and discernible, as recognizable as the bright strings of a violin or the deep thrum of a cello in a symphony.

Perception is the act of the mind's chef, blending these individual notes to create a gourmet experience, a melody that is greater than the sum of its parts. Just as your brain weaves together the symphony's instruments into music that stirs the soul, it combines the singular sensations of lemon zest, cinnamon, and cream into the delectable whole of a dessert. It's this orchestra of senses, fine-tuned and conducted by the brain, that allows you to not only identify each flavor or instrument but to revel in the entire culinary or musical masterpiece as one complete, unforgettable perception.

Picture yourself sitting in a bustling café. Your brain is like a skilled conductor, seamlessly merging the aroma of coffee, the warmth of sunlight streaming through the window, the softness of the chair, and the chatter around you into a single, coherent experience. First, receptors specific to each sense capture their respective stimuli - olfactory cells react to the coffee's scent, the skin registers the sun's gentle heat, the ears catch snippets of conversation, and the eyes take in the room's ambiance.

These receptors send their raw data as electrical impulses through dedicated neural pathways, akin to couriers rushing to deliver important messages. The impulses travel to the brain's relay station, the thalamus, which then redirects them to specialized areas in the cerebral cortex designed to process each sense individually.

But the magic happens when the cortex blends these individual sensory inputs, crafting them into a singular, multidimensional experience. Neurons from these different sensory areas communicate and cross-reference information, fact-checking and filling in gaps to create a perception that makes logical sense. Like a maestro ensuring harmony among the ensemble's instruments, the brain synchronizes these sensory cues.

This process allows one to appreciate the café not just as isolated sights, sounds, tastes, and touches, but as a vibrant setting where one feels immersed. The brain's ability to synthesize these sensations into a unitary perception is what enables one to live in and navigate a world that's rich with complexity and nuance.

Let's take a deeper look at how the cerebral cortex turns a multitude of signals into a singular, vivid perception. Once sensory stimuli are gathered by receptors, they transform into electrical pulses that embark on a journey to the brain. Entering the cortex, these impulses are like raw ingredients delivered fresh to the kitchen of a skilled chef.

Upon arrival at the primary sensory cortices, these signals are like individual food items, each identified and prepped. The thalamus is responsible for sending these signals to the correct cortical address. Like a doorkeeper, it organizes and directs traffic, ensuring that messages about sight, sound, touch, taste, and smell reach the appropriate corner of the brain for processing.

Neurotransmitters are the vehicles of communication in this bustling hub, shuttling information across synapses—tiny gaps between neurons. The synapses adjust their strength in response to activity, a property known as synaptic plasticity—akin to kitchen staff sharpening their skills, becoming faster and more efficient with practice.

But what turns individual sensations into perception is the brain's remarkable ability to cross-reference these signals. This happens in the cortical association areas, where the assorted sensory data is mixed and matched, evaluated, and finally blended. Imagine these areas as the chef's station, where flavors are combined with expertise to assemble a dish whose taste transcends the sum of its singular components.

The prefrontal cortex acts as the head chef, adding seasoning—context—to the dish. It sorts through memories and current sensory data to anticipate the diner's preferences and expectations, finishing the plate to perfection.

The resulting dish—the unified perception—is a masterpiece of the mind's inner kitchen. For example, imagine the simple act of enjoying a cup of coffee. Heat gently radiates through the mug, the rich aroma wafts up, the bitter taste mingles with sweetness on the tongue, and soft background music caresses the ears. These separate sensations meld into the holistic experience of 'having coffee,' courtesy of the brain's complex interplay of sensory integration. Through this lens, one can begin to appreciate the brain's prowess as both a biochemical marvel and an artist crafting the continuous experience of existence.

Imagine strolling through an art gallery with a friend. You both set eyes on the same vibrant painting, but while you perceive a tranquil ocean at dusk, your friend envisions a turbulent sea under a stormy sky. This discrepancy isn't about eyesight; it's about perception, about how each brain interprets the same swirls of blue and strokes of grey. Perception is individual, as uniquely crafted as one's fingerprint.

Now, think of a song playing in a crowded room: some might hone in on the lyrics, connecting with the story they tell, while others may find their bodies swaying to the rhythm, captivated by the beat. These sensory signals – sound waves – are identical for all ears in the room, yet the brain of each listener tunes into different aspects based on past experiences, mood, or even what they're focusing on.

These differences are a window into the infinite interpretations of reality that human brains can conjure. It's like sampling a complex dish; each spice or ingredient might stand out differently to different palates, drawing from a repository of culinary memories and preferences. That's the beauty and complexity of human perception – it transforms objective sensory information into a rich tapestry of subjective experiences, painting a world where each of us lives out our own unique version of reality.

Here is the breakdown of the factors that contribute to the unique tapestry of individual perception:

- Reception of Sensory Input

- Sensors across the body gather data: eyes for light, ears for sound, skin for touch, tongue for taste, and nose for smell.

- Sensory memory briefly holds onto these impressions, like an artist's preliminary sketches before the full picture is developed.

- Directed Attention

- The brain allocates focus, directing it towards certain stimuli while filtering out others, similar to a photographer choosing what to center in the frame.

- Attention sharpens perception of what's deemed important and dims the background, just as a spotlight illuminates an actor on stage while casting shadows elsewhere.

- Accessing Memory

- Long-term memory acts as a reference library, with the brain comparing new sensory data against stored information to find matches, akin to a musician tuned to a specific note.

- These memories help identify familiar patterns or trigger associative responses like the warmth of a melody reminding someone of a past summer day.

- Emotional Influence

- Emotional states tint perception, much as colored lenses can alter the appearance of a scene.

- Joy can make colors seem brighter, sorrow can mute sounds, and anxiety might sharpen scents, all influencing how one experiences the same situation differently.

- Cultural and Personal Preferences

- Personal history and cultural context shape the interpretation of stimuli, influencing preferences in a similar way to how tastes in food develop: spicy cuisine may appeal to some while repelling others.

- Music perception can vary, with certain rhythms resonating in one culture while another finds melody to be the soul of musical expression.

Drawing on these analogies captures the complexity behind each personal encounter with the world. Like painters selecting colors from their palette or musicians composing a piece, individuals create their own perception, crafting a distinct view of reality through the intertwining of sensation, memory, emotion, and culture. This intricate weave of influences ensures that no two people experience the world in the exact same way, reinforcing the splendor found in the diversity of human perception.

Helen Keller's extraordinary journey illuminates the remarkable adaptability of the human sensory system. Although blind and deaf from a young age, she learned to experience the world through her remaining senses, demonstrating how perception is not solely dependent on the full spectrum of sensory input. With the determined help of her instructor, Anne Sullivan, Keller decoded the tactile world, transforming touch into a vessel for knowledge and communication.

Modern neuroscientists continue to unravel the mysteries that Keller lived. Their research, probing the depths of how sensory data is translated into perception, echoes Keller's lived experience of learning and adaptation. Neuroplasticity, the brain's ability to reorganize itself by forming new neural connections, is no longer just a concept but an observable phenomenon, one that Helen Keller's life personified before the term was even coined.

The work of these scientists offers insight into how Keller's brain might have compensated for her sensory losses, repurposing areas typically involved in sight and hearing to enhance her sense of touch and smell. This phenomenon sheds light on the flexible nature of the brain and its capacity to find new ways to perceive the environment.

Presented in a manner akin to a guide leading a novice through a complex landscape, the stories of Keller and contemporary neuroscientists serve as landmarks, demonstrating the expansive capabilities of the human brain. They toast to the spirit of inquiry and resilience, joining past experiences with present-day science to map the terrain of perception—a topic as dense as it is fascinating, articulated here to foster understanding and ignite curiosity.

Dive into the world of virtual reality and haptic technology, where the boundary between the natural and artificial blurs. It's like slipping on a pair of gloves that can replicate the sensation of touch or donning goggles that transport you to another world. These devices are master mimics, sending signals designed to interact with your sensory systems as if they were the real thing.

In VR, visual and auditory stimuli are crafted with such precision that your brain is convinced you're gazing at a sunlit forest or hearing the rush of a waterfall. It's akin to stepping onto a stage where the lighting and sound are so expertly managed that for a moment, the illusion becomes your reality. The haptic feedback in a wearable device simulates the pressure and texture of objects, like an invisible sculptor guiding your hands to feel shapes and materials that aren't there.

This technology taps into the body's natural sensory processing pathways, mirroring the way real stimuli are interpreted by the brain. It's as if your senses have been given a script to perform a play, and the technology cues every line and action, seeking to make the scene as believable as possible. Through these simulated experiences, the marriage of human perception and artificial input is celebrated, demonstrating both the sophistication of human sensory systems and the ingenuity of technology designed to engage with them.

Let's take a deeper look at the intricate dance between virtual reality and haptic technology and the human sensory system. Virtual reality (VR) headsets are designed to create an illusion of depth and space, tricking the visual system into perceiving digital images as part of the surrounding environment. They use finely tuned display technology that includes high-resolution screens close to the eyes, with each eye viewing a slightly different angle, replicating the way the human visual field works to create stereoscopic depth. Meanwhile, audio in VR makes use of binaural or 3D sound, using stereo speakers or headphones to mimic the way sounds reach us in different ways, playing on the natural acoustic cues we use to locate and identify sounds in space.

Haptic technology simulates the sense of touch through a variety of feedback mechanisms. Vibrations, delivered by small motors called actuators, mimic the sensation of movement or impact. Other haptic devices can exert force or resistance against the user's movements, providing a sense of solidity and weight akin to lifting or pushing real objects. These stimuli engage the same neural pathways that interpret touch, temperature, and proprioception - the sense of bodily position and movement in space.

When VR and haptic inputs flood the senses, the brain processes these artificial cues similarly to real-world stimuli. Neural signals zip from sensory receptors to the brain's thalamus and on to the primary sensory cortices. The thalamus acts as a relay station, vetting and forwarding sensory data. The sensory cortices, located in various crannies of the brain's landscape, analyze these inputs, and in collaboration with the prefrontal cortex, piece together a coherent experience.

To enhance immersion and curtail the dissonance between real and virtual experiences, these technologies employ methods such as predictive tracking and adaptive interfaces. Predictive tracking anticipates user movements to reduce lag, while adaptive interfaces adjust in real-time to user interactions. This ensures that the simulated world responds in ways that match physical expectations, stifling any jarring mismatch that would break the illusion.

VR and haptic engineers tirelessly refine their crafts to narrow the rift between simulation and reality, mirroring natural sensory experiences as closely as possible. By understanding and harnessing the detailed workings of the human senses, these technologies act as a bridge, extending the capacity for experience beyond the confines of physical reality, into realms only limited by imagination.

As you step away from this exploration of senses, take a moment to marvel at the symphony your brain conducts every second of every day. From the gentlest whisper of wind to the kaleidoscope of colors at sunset, your sensory experiences are nothing short of a masterpiece painted on the canvas of consciousness. This journey—from raw, chaotic data to a perception as tailored and personal as a signature—reveals the silent choreography that unfolds within. It underscores how a simple touch, a single note, a lone fragrance, becomes part of a grand narrative stitched together with invisible threads by the mind's own hand.

Whether it's the memory-laden scent of a grandmother's perfume or the intricate flavors of a well-crafted meal, pause to appreciate these everyday marvels. Each sensation, no matter how fleeting, is a note in the opus of perception. And just as the depth of a novel lies in the subtlety of its prose, so too does the depth of reality emerge from these sensory nuances. This understanding reminds you to revel in the sensory world and its complex beauty, acknowledging the brilliant biological tech that makes it all possible. Breathe it in. Listen closely. Look around. Touch. Taste. Perceive. Embrace the richness of the sensory gifts that allow you to navigate and find joy in this complex, vibrant world.

MEMORY AND LEARNING

We now enter the awe-inspiring arena of the human brain, where each day is a new act in the grand performance of memory formation. It's where moments transform into memories, and experiences forge the foundations of knowledge. As this story unfolds, readers will discover the brain's ability to capture the fleeting twirls of now and craft them into the enduring sculpture of recall. Think of this process like a gardener who tends not only to the blooming flowers of the present but also to the seeds that will sprout into the rich flora of tomorrow's recollection.

This chapter invites readers on a relatable journey through the processes that dictate how one remembers a dear friend's laugh, learns the steps to a favorite dance, or recalls the comforting scent of home. It's a journey that meticulously breaks down the mechanics of memory, from the birth of a neural impulse to the complex networks that store our life's tales.

As a guide through this labyrinth, the narrative steers clear of technical jargon, opting instead for clear signposts and analogies that resonate with lived experience. This voyage is as much about marveling in the capacity of the brain as it is about understanding it; it's a venture into the marvel of memory, a tour of an inner cosmos as boundless as the stars above. Ready your senses, for this adventure into memory and learning is about to begin — an exploration designed not just to inform but to fascinate and illuminate the extraordinary capabilities nested within everyone's grey matter.

Imagine for a moment that short-term memory is like the basket of a bicycle — it's handy for the errands of the now, holding onto groceries just long enough to get them home. It's there for the temporary tasks, a fleeting storage space where the names of new acquaintances hover for a moment before slipping away if not transferred to a safer place. By contrast, long-term memory is the robust cellar, deep and cool, where fine wines are stocked, aging and absorbing complexity. It's a place for safekeeping life's vintage moments and intricate learnings, ensuring that they can be savored and recalled years down the line.

Short-term memory processes the stream of information just as the basket carries fresh bread; both serve immediate purposes but are not meant for holding onto their contents indefinitely. Long-term memory, with its elaborate racks and rows akin to the cellar's shelves, is more discerning — it chooses what to keep based on meaning, repetition, or emotional impact.

This understanding of short-term and long-term memory provides not just a glimpse into their unique functions but emphasizes their integral roles. Whether grasping a phone number long enough to dial (like keeping ahold of a delivery slip until the package is in hand) or embedding a cherished memory of a family trip (similar to storing away one's wedding album), each type of memory enriches life, granting you the tools for the present moment and the treasures of a lifetime.

Here is the breakdown on the neural mechanisms of memory:

- Basic Structure of a Neuron

- Dendrites: Branch-like extensions that receive signals from other neurons, comparable to the antennae of a radio, picking up signals to be processed.

- Axon: A long, thin projection that transmits signals away from the neuron's cell body, similar to a fiber optic cable carrying data.

- Synapses: Microscopic junctions where neurons communicate with each other, functioning like micro-ports where information is transferred from ship to dock.

- Synaptic Plasticity

- Neurotransmitters: Chemicals released by neurons that cross synapses to communicate with receptor sites on another neuron, akin to keys that unlock doors to signal transmission.

- Receptor Sites: Proteins on the neuron's surface that bind to neurotransmitters, much like a lock waiting for the right key.

- Memory Trace Strengthening: Repeated activation of the same synaptic pathways enhances the connection, building a stronger memory, much as a well-trodden path in the woods becomes more defined with use.

- Brain Regions Involved in Memory

- Hippocampus: The hub for long-term memory and spatial navigation, serving as a central processor that turns short-term memory into long-term storage, similar to a processing plant that packages goods for shipment.

- Prefrontal Cortex: Responsible for working memory, which holds information temporarily for cognitive tasks, comparable to a computer's RAM that temporarily holds data for ongoing processes.

- Neurogenesis and Its Role in Learning and Memory

- Formation of New Neurons: Occurs in certain brain regions and is crucial for adaptability and learning, like planting new seeds to cultivate a garden of knowledge.

- Importance in Memory Consolidation: New neurons contribute to the stability and durability of memories, ensuring that they persist, in much the same way adding new materials can fortify a building.

- Real-life Parallels

- Strengthening of Synaptic Connections: This is analogous to reinforcing the structure of a bridge, so it can carry more traffic and withstand the test of time, signifying stronger, more resilient memories.

By presenting these intricate processes with precise, relatable terms and clear analogies — comparing parts of neurons to familiar objects or comparing brain structures to components in a city — one can better grasp the sophistication of memory formation. These insights into memory's building blocks reveal a world within, as complex and miraculous as any metropolis or ecosystem, highlighting the grand architecture of the human brain.

At the heart of learning is a remarkable network of neurons, which are like the tiny electric wires of the brain. These neurons talk to each other at places called synapses, which function as miniature conversation points where messages hop from one neuron to the next. Think of synapses as whispering galleries, where a soft message travels clearly across a domed ceiling from speaker to listener.

Now, when something new is learned, like the steps of a dance, it's not just about one whisper getting passed along. It involves a full chorus of neural conversations, with the synapses playing their part to strengthen the message, reinforcing the dance steps every time they're practiced. It's much like repeating a phone number to remember it; each time the number is recited, the memory becomes firmer.

Key areas in the brain, such as the hippocampus, are like the conductors of this neural orchestra. Located deep within the brain's folds, the hippocampus gives the cue for when and how these neural messages should be amplified, akin to a librarian who not only stores the books carefully but also knows exactly which book to fetch from the shelf when needed.

In summary, neurons form the connections, synapses pass the baton of information, and the hippocampus ensures that the information is not just stored but also organized. It's a complex yet synchronized activity that takes the 'new' and turns it into 'known', allowing an individual to perform a combination of dance moves or solve a challenging puzzle. Understanding these roles provides a clear look at the rich, dynamic process of learning, revealing just how wondrous the brain's capabilities truly are.

Learning is fundamentally about making and strengthening connections—a biological process involving neurons and synapses. Here's a simple breakdown of how these elements work together:

The process starts with an action potential, an electrical surge that travels down a neuron's axon, much like a wave rushing along a water slide. When this wave reaches the end of the line, it prompts the neuron to release specific chemicals called neurotransmitters. These chemicals are the couriers, carrying messages across the synapse, the tiny space between neurons. They bridge this gap by binding to receptor sites on the neighboring neuron, much like a key fitting into a lock, allowing the message to continue its journey.

There's a whole ensemble of neurotransmitters, each playing a unique role in the learning orchestra. Dopamine, for example, is like the reward sticker one gets for a job well done, reinforcing behaviors that are beneficial and creating a sense of pleasure. It's particularly important in reward-based learning, contributing to the motivation to repeat actions that result in positive outcomes.

Then there's long-term potentiation (LTP), a vital cellular process for learning and memory. It's akin to strengthening a muscle through exercise—the more a synaptic connection is used, the stronger it becomes. In LTP, repeated stimulation of synapses leads to increased sensitivity and structural changes that enhance signal transmission. It's like paving a dirt path into a wide road due to frequent travel, making future journeys along this route faster and more efficient.

The hippocampus is a star player in transforming short-term memories into long-term storage. Imagine you're copying important files from your computer's temporary memory to an external hard drive for safekeeping—that's akin to the hippocampus's role. It selects which short-term memories are worth saving and then steadily incorporates them into the long-term memory bank.

Finally, neuroplasticity is the brain's way of adapting to new challenges and information. It allows the brain's neural networks to remodel themselves, like a railway system that's constantly updating and adding new tracks to improve connectivity and efficiency. Neuroplasticity ensures that learning is not a rigid process but a dynamic one, capable of reshaping the brain's pathways to accommodate new knowledge and skills.

Together, these components form the incredible system that enables humans to learn—from understanding complex concepts to mastering new abilities. The story of learning is written across the neurons, through the synapses, and within the brain's very structure, marking a journey of continual growth and adaptation.

Think of a momentary experience as a beautiful vista that you happen upon during a hike. In an instant, your brain is like a camera snapping a photo of this scene. The light, the colors, and the depth all captured in a fraction of a second. This is the moment of encoding, where your senses are the camera's lens, focusing the details into a crisp image to be saved.

Just like a camera stores the photograph on a memory card, your brain begins the process of committing this experience to memory, storing it on the neural pathways. The more significant the moment or the more often you 'revisit' it by thinking about it, the stronger the memory becomes—a bit like editing and saving the photo in higher quality or with more vivid colors each time you review it.

The photograph—now a saved memory—can be retrieved and viewed anytime, bringing back the rush of emotions and the vividness of the first encounter. This is recall, the act of looking back at the mental photograph your brain has taken and kept safe amongst the album of your other cherished memories.

Through this simple but powerful analogy, we can appreciate how our brain's memory works in harmony with our experiences to preserve moments, both monumental and minute, allowing us to revisit and relive them, adding depth and details to the narrative of our lives.

Let's take a deeper look at how our brains capture and keep hold of our experiences. When a moment strikes the senses, it's like a camera taking in light to create an image; our sensory receptors are gathering information to send to the brain where it creates a 'mental snapshot' or an initial memory.

Starting with sight, sound, touch, taste, or smell, these inputs trigger electrical impulses that travel along the neurons, like speed dialing a message through phone lines. These impulses reach synapses, the meeting points of neurons, like relay racers passing the baton. Here, neurotransmitters, the chemical messengers of the brain, are released and bind to receptors on the receiving neuron. This is like inserting a key into a lock, opening the door for the next message to come through.

The strength of these connections, which form our memories, is influenced by the repetition of experience—rehearsing a fact or skill—and the emotional significance of the event. It's like carving a path through a forest; the more you travel it, or the more meaningful the destination, the clearer and more durable the path becomes.

Beyond the paths themselves, the hippocampus acts as a sort of sorting center, deciding which memories are kept and which are discarded. Important or frequently revisited memories are transferred from short-term holding into a more permanent storage, like moving files from a computer's desktop to a hard drive.

Long-term memories are stored through changes in synaptic strength and sometimes by the formation of new synaptic connections—a process known as long-term potentiation. When we need to recall a memory, these synapses are reactivated, allowing us to revisit past experiences. This is like having a well-organized photo album that you can flip open to any picture you want to see again.

However, every time we bring up a memory, we might edit it slightly based on new information, perspectives, or even the mood we're in during the recall—a bit like using a photo editing software to tweak an image. Over time, this can alter the 'original photo' stored in our memory.

By understanding each of these steps—from the 'camera' capturing the scene, through the 'relay race' at the synapses, to the 'photo album' in our long-term storage—we gain not just insight into the marvel of memory but also an appreciation of the brain's complexity and adaptability in preserving our personal narratives.

Consider the brain's memory system as the ultimate sophisticated data archiving solution, far beyond the capacity of even the most advanced computer storage. The brain stores away memories like a cloud service meticulously organizes data across multiple servers. A vast network of neurons acts as this storage matrix, each synaptic connection a hyperlink to a stored file, accessible within milliseconds.

The brain's remarkable aptitude for retaining information can be seen in individuals with extraordinary memory capacities. Take, for instance, dominators of memory championships, who, like high-powered search engines, retrieve the exact data – may it be sequences of numbers, decks of cards, or lists of names – with jaw-dropping accuracy and speed. Or consider historical prodigies like the late Kim Peek, who served as the inspiration for the film "Rain Man," and could recall the contents of over 12,000 books, a feat akin to a supercomputer's ability to process and recollect vast amounts of information on command.

These real-world examples are testaments to the brain's storage system – a complex yet flawlessly operating archive that sorts, files, and recalls information. From the intricacies of a savant's recall to the everyday ability to remember a beloved melody, the mechanisms underpinning the brain's memory capabilities invite admiration and a deeper curiosity about this incredible organic data center we all possess.

Let's dissect the parallels between the brain's memory system and a computer's data processing functionalities to illuminate the seamless and sophisticated nature of our cognitive hardware.

The encoding process in the brain, akin to data entry in a computer, commences when you experience something new. Your sensory systems gather information and convert it into neural signals, like typing up a document and saving it to your computer's drive. These signals travel to the hippocampus, which plays a key role in deciding what gets saved and what's discarded — not unlike a computer's central processor deciding which data is crucial enough to keep.

The consolidation phase is the brain's version of defragmenting. Over time, particularly during sleep, the brain sorts these neural signals, tying them into existing knowledge networks, fortifying the connections like a computer reorganizing its files for quicker access and more efficient storage. It's an optimization process that ensures information is filed away in a logical, retrievable format.

Retrieval of memories happens when you recall information — it's like clicking on a file and opening it on your computer screen. The brain scans through its synapses looking for the right connection to bring forth the memory. Sometimes, memories are updated with new information or corrected; think of it as running an update or patch on your software, ensuring the system performance is as accurate and current as possible.

Talking about memory capacity and retrieval speed, your brain doesn't quite run out of space like a physical hard drive because it's constantly rewiring and reconfiguring — a dynamic storage expansion. However, how swiftly you can recall information depends on how frequently and recently you've accessed that memory, similar to how a computer's RAM impacts its ability to quickly fetch and process data.

Lastly, synaptic plasticity — the brain's ability to strengthen connections based on frequency of use — is much like a system upgrade in a computer. Just as new software can increase a computer's efficiency, repetitive learning and practice enhance the synaptic pathways, improving your ability to remember and perform tasks more adeptly.

By drawing these parallels, the complexities of the brain's memory processes can be matched step-by-step to a realm most are familiar with — computational data processing. It's an interplay of remarkable biological engineering that underlies the infinite capabilities of human memory and learning.

Think of memory retrieval as being as streamlined as typing a query into a search engine and hitting 'Enter.' Just as you would pull up information on the internet in seconds, the brain can access a memory with ease, instantly providing you with the 'search results' of your inner database. This process is like the mental equivalent of autocomplete; as soon as you start to think of an event or a fact, your mind races ahead, filling in the details and presenting the full picture.

Experts who master memory recall — like the memory champions who astonish us with their ability to memorize decks of cards or endless strings of numbers — perfect this retrieval process. They're akin to search engine optimization specialists; they've learned precisely how to index and recall data efficiently. Through techniques such as the method of loci, they create a 'mind palace,' placing memories in imagined locations, then navigating through this palace with such familiarity that they can instantly 'click' on the right memory, just as you would click on a search result to get detailed information.

Just as certain search terms bring up the most relevant web pages, these memory athletes use specific mental cues to hone in on the memory they need, bypassing irrelevant information to focus on the prize. This dazzling ability underscores that with practice, our natural capacity for memory retrieval can become as effective and as rapid as our favorite online search tool, proving that the brain, indeed, is the most powerful information processor we have at our fingertips.

Let's take a deeper look at the delicate dance of memory retrieval, a feat that our brains perform with spectacular complexity and finesse:

During recall, the hippocampus and cortex collaborate closely. Picture the hippocampus as a director, cueing the brain's various departments – the sensory cortex, language areas, emotion centers – to piece together the fragments of a memory. This director doesn't work alone; it relies on the cortex to flesh out the full storyline, pulling from the rich tapestries of detail stored across various brain regions.

The neural events that reconstruct a memory from an initial thought or cue can be likened to a domino effect. The spark of recall – perhaps a smell or a word – sends electrical impulses cascading through neural pathways, reactivating the patterns of activity that were initially recorded. Like finding the first piece in a puzzle, this cue helps the brain assemble the rest of the memory's picture.

Neurotransmitters, the brain's chemical messengers, either grease the wheels of this process or apply the brakes. They manage the flow of information by regulating the strength of the synaptic connections – dopamine might speed things up, aiding quick retrieval, while substances produced under stress might dampen the activity, making recollection foggy.

Memory retrieval is sensitive to context, and external factors such as stress levels, the environment, and one's emotional state can color the process. A stress-free mind in a familiar setting might recall with sharp accuracy, much like a well-tuned instrument playing a clear note. But under anxiety, the same mind might grapple with memory, like trying to tune into a radio station amidst static.

Memory champions use strategies such as the method of loci, which involve assigning information to imaginary locations, to enhance recall. By navigating through their 'mind palace,' each 'room' or locus conjures vivid imagery tied to specific data, indexing memories in a spatial, organized fashion for swift, precise retrieval. It's a method that harnesses the brain's innate spatial memory skills, transforming them into a powerful tool for recalling even the most complex information.

Peering into these aspects of memory retrieval uncovers a world of mental acrobatics that rivals any computer in speed and efficiency. It's a reminder of the brain's immense power to capture, hold, and revive the countless experiences that shape the narrative of our lives.

Just as hitting the 'save' button on a document before closing your laptop ensures your work isn't lost, sleep serves a similar critical role in safeguarding our memories. Consider sleep as the brain's own time for file management and system optimization, running a nightly backup of the day's experiences. As we sleep, the brain isn't just resting; it's busy transferring short-term memories, like a day's conversations or lessons learned, into more durable, long-term storage.

This process is much like saving files to a hard drive, where they can be neatly stored and easily retrieved later. During deep sleep stages, our brains are hard at work solidifying connections between neurons, a process known as 'synaptic consolidation.' This is the nitty-gritty work of turning today's experiences into lasting knowledge—like turning scribbled notes into a finalized report.

Moreover, much like a computer's disk defragmenter reorganizes data for more efficient operation, the brain uses sleep to reorganize and strengthen memories. It pares down unnecessary 'mental clutter' from the day and reinforces the pathways of what matters, ensuring that crucial information is anchored securely and the cognitive system runs smoothly the next day.

Remarkably, during these off-hours, our brain is also running diagnostics, identifying errors from the day's learning and making corrections where necessary. It's a nightly tune-up that primes us for clarity and focus, readying us to build upon yesterday's foundations with today's new inputs. Sleep, therefore, isn't just a pause from wakefulness; it's an active, dynamic period where the brain's maintenance ensures our cognitive processes run smoothly for the long term.

Sleep unfolds in stages, with each one acting like a different program designed to run at specific times to maintain the brain's memory system. Like a computer scheduled to perform various tasks like virus scanning and file backup at different hours, the brain cycles through REM (rapid eye movement) sleep and several stages of non-REM sleep, each serving unique functions in memory consolidation. During deep non-REM sleep, for example, the brain may be 'saving' procedural memories, such as how to ride a bike, akin to a computer transferring data to a secure drive for long-term file retention.

As a person sleeps, the brain undergoes various biochemical changes pivotal for memory consolidation. The release of neurotransmitters during sleep can be akin to a computer sending commands to ensure data is appropriately processed and stored. Neurotransmitters such as acetylcholine and norepinephrine fluctuate in levels, orchestrating the strengthening or weakening of synapses, very much like how a computer program determines when to move or delete certain files based on predefined algorithms.

Brainwave patterns play a significant role here, with changes in their frequency and amplitude during different sleep stages acting to reinforce synaptic connections. It's like comparing the ebb and flow of internet bandwidth; certain times are primed for downloading large files, much like certain sleep waves are primed for solidifying crucial information.

Error correction in this context is the brain's method of aligning memories with reality, much like a software program cross-referencing data to fix errors during a system backup. During sleep, particularly during REM stages, the brain evaluates memories, consolidating them while correcting mistakes. It's as if the brain is running a nocturnal audit, ensuring that the 'documents' of the day's experiences are accurate and coherent.

Lastly, neuroplastic changes that occur during sleep can be thought of as updates to the cognitive 'software,' improving performance and memory precision. Like a computer receiving new programming that allows it to run more efficiently, the brain rewires itself to be faster and more accurate in retrieving memories.

This complex interplay of sleep stages, biochemical activity, brainwave patterns, and error correction contributes to a proficient system that, under ideal circumstances, enables a state-of-the-art 'software'—the human mind—to learn, adapt, and recall with exceptional skill. These processes remind us that a good night's sleep is not just restorative; it's a critical period where the brain is actively fine-tuning its ability to remember and learn.

As the final notes of this melody on memory and learning fade, it's worth pausing to appreciate the symphony of processes that so intricately shape our identities. Each piece of knowledge learned, every memory etched into the canvas of our minds, serves as a brushstroke in the portrait of who we are. Our capacities to remember the past, to learn from experience, indeed to dream of the future, all stem from the boundless complexities of the brain.

Think of the first book that truly touched you, the aroma that takes you back to your grandmother's kitchen, or the muscle memory that kicks in when you ride a bike. These are not mere fragments of data; they are the construction blocks of the self, the very essence of persona. The artistry with which our minds sculpt these moments into coherent narratives is nothing short of miraculous.

Embarking on a journey through the landscapes of the mind reveals not only how memories are made and kept but also their influence on every step we take. It brings a profound respect for the brain's capabilities—the capacity to adapt, the gift of reflection, the constant reshaping of our knowledge base, all contribute to a richly lived life. Reflecting on this intricate tapestry fosters a sense of wonder and, perhaps, a deeper acknowledgement of the brain as the most magnificent storyteller, tirelessly weaving the epic saga that is the human experience.

CONSCIOUSNESS AND COGNITION

It's time to step into the journey within consciousness and cognition, a path that leads straight to the core of what it means to be human. This is an adventure into the very essence of thought and awareness, a venture that's both as intimate as the echo of your own voice and as vast as the universe of your mind. Imagine piecing together a puzzle that reveals not just images, but entire worlds—worlds of memory, decision-making, and perception. Here, you'll unravel the threads of awareness that weave through every moment awake or in dreams. You won't find industry jargon or impenetrable concepts on these pages. Instead, what awaits is a clear, methodical guide through the intricate landscapes of the mind. With each step, every complex idea will unfold into relatable insights, allowing a deep connection to the material that resonates with your lived experience. So pause, take a breath, and prepare to engage with the mysteries of consciousness—the silent narrator of your life's story—and cognition, the artist that paints your reality with every perception and choice.

Understanding consciousness and cognition is akin to exploring how a silent narrator within us perceives the story of our lives while an unseen artist paints our perceptions and choices, layer by layer. Consciousness is the light that illuminates our internal world, the profound state of awareness that allows us to experience sensation, to reflect upon our own existence, and to savor each moment's taste. It's the gentle voice of the self, always present, always observing – as private and personal as the warmth of the sun on our face or the chill of a breeze.

Cognition, meanwhile, is our mind's toolbox; it's the set of mental tools we use to navigate the world. It's how we remember where we left our keys, it's the decision-making process we go through when we stand before a crossroad, and it's the language we use to express our deepest feelings and smallest whims. It's also the ability to learn from yesterday's mistakes, to plan for tomorrow's ventures, to solve the puzzle of a broken appliance, and to create from the heart and mind.

These two – consciousness and cognition – are what enable us to experience life in vibrant color and high definition. Together, they form the central pillars upon which our thoughts, memories, feelings, and volitions rest. This narrative will take you through their mechanics in clear, everyday language, illustrating their operation with scenarios and examples that shine light on their workings within your own daily experience. From the minute you awaken to a new day to the moment you drift off to sleep, consciousness and cognition are the silent partners guiding your journey through every waking moment.

Let's take a deeper look at the tapestry of processes crafting our conscious experience and the neural ballet that is cognition. Consciousness sprouts from a symphony of brain activities, with regions like the prefrontal cortex, which is like the conductor of a grand orchestra, setting the tempo and pulling the strings of our awareness. The thalamus acts as the central hub, much like a busy train station, channeling sensory and motor signals to various destinations within the brain, contributing to our perception of being in the 'here and now'.

Cognition unfolds through the firing of neurons, creating pathways as intricate as the streets of a teeming metropolis. These neural pathways, formed and reinforced by connections called synapses, are the highways of thought, memory, and decision-making. Imagining the hippocampus as an archive or database allows us to understand its critical role in memory retention, meticulously cataloging and storing memories for later retrieval.

Neurotransmitters are the messengers whisking along these highways, the texts and emails keeping the city running. They facilitate cognitive functions, with serotonin adjusting mood, dopamine spurring motivation, and acetylcholine fine-tuning attention and learning. A shift in brain chemistry, like an alteration in a city's communication network, can cause significant changes in consciousness, such as the bright lift of alertness or the fog of confusion.

Analogy serves as a bridge over these complex biological riverways. Consider the brain's signal transmission comparable to the internet's data streaming, where information is uploaded and downloaded with high-speed precision, reflecting the swift interchange of thoughts and responses marking our every waking moment. This narrative isn't just about the brain's circuitry; it's a guide to the living, breathing neural ecosystem that makes us who we are, capable of reflecting on the past and imagining futures yet to unfurl. The exploration into consciousness and cognition is thus a journey not just into the brain's matter but into the heart of human existence.

Picture consciousness as the silent, ever-awake storyteller of your mind, weaving together a narrative from countless sensory inputs—much like how a director pieces together a scene in a film, selecting from an array of clips to craft a seamless whole. It is an inner spotlight, one that shines on the thoughts and feelings playing across the stage of your mind. In the theater of your daily experiences, consciousness is both the audience and the stage, simultaneously observing and presenting the unfolding script of your life.

Cognition, on the other hand, is the behind-the-scenes crew, the scriptwriters, set designers, and sound engineers working diligently to support the main attraction. When you solve a problem at work, it's like being a detective in a crime drama piecing together clues; each cognitive function contributing to crack the case. When you learn a new skill, imagine it as constructing a bridge in a city, opening new pathways and connections that facilitate travel and communication.

Attention, a crucial aspect of cognition, can be likened to a camera's focus—zooming in on a subject, blurring the background, and capturing the finer details that otherwise might be overlooked. Just as a photographer adjusts the lens to clarify an image, your mind hones in on the essential, allowing you to concentrate amid the bustle and distractions that surround you.

These analogies illuminate the profound yet subtle dance between consciousness and cognition, inviting you to appreciate the elegant complexity of these concepts. Understanding them is like tuning an instrument; the more finely adjusted it is, the more harmoniously it plays, enriching the music of our lives with melodies of clarity and insight.

Here is the breakdown on the key components that orchestrate the phenomena of consciousness and the operations of cognitive tasks:

- Brain Regions and Consciousness:

  - Prefrontal cortex:

    - Orchestrates thoughts and decisions

    - Focuses attention and filters distractions

- Thalamus:

  - Relays and prioritizes sensory information

  - Acts as a hub for alertness and awareness

 - Amygdala:

  - Processes and encodes emotional responses

  - Influences the emotional coloring of memories

- Cognitive Function Steps:

 - Encoding:

  - Translates experiences into neural language

  - Tags and categorizes information for later access

 - Consolidation:

  - Solidifies temporary neural connections into longer-term synapses

  - Weaves new information into the existing memory network

 - Retrieval:

  - Searches and locates stored memories within the neural network

- Reactivates the neural pattern corresponding to the memory
- Influence of Neurotransmitters:
  - Dopamine:
    - Boosts the drive and desire to pursue goals
    - Sharpens focus and enables the reinforcement of vital experiences
  - Serotonin:
    - Moderates mood swings, leading to emotional stability
    - Influences social behavior and risk-taking tendencies
  - Acetylcholine:
    - Enhances the ability to learn new information and tasks
    - Strengthens attentional focus, particularly during learning
- Impact of External Factors:
  - Sleep:
    - Facilitates the pruning of unnecessary synapses and the reinforcement of important ones
    - Allows the brain to 'clean-house', removing waste byproducts that

accumulate during wakefulness

  - Stress:

    - Elevates hormones like cortisol, which can cloud thinking and impair judgment

    - May induce a 'fight or flight' response, narrowing focus but limiting broader cognitive capacity

  - Environment:

    - Provides contextual cues that can trigger memory recall

    - May increase cognitive load, challenging the brain's processing capacity

Each bullet paints a piece in the elaborate jigsaw puzzle of mental life. Like a maestro steering through a complex symphony, the dynamics of these components come together to create the rich tapestry of conscious experience and cognitive function. This guide walks you through a field that's constantly evolving, as each discovery brings us closer to answering age-old questions about the mind's capabilities and limitations – an endeavor that's as fascinating as the enigma of the cosmos above or the subatomic worlds below.

Consider the brain — a masterful and intricate command center. It's a bit like a bustling airport, where thoughts, memories, and emotions are passengers, traveling along neural pathways that crisscross like air traffic routes. Each region is specialized, akin to the different terminals dedicated to arrivals, departures, and cargo. The prefrontal cortex is the executive lounge where plans are thought out and decisions are made with a calm overview of the busy runways. Deep within, lies the thalamus, a central hub directing sensory signals to their final destinations with the precision of an air traffic controller. The amygdala is like the airport security — keeping a watchful eye on the feelings rushing through, ensuring emotional responses are appropriate to their context.

These structures work in concert, their activities synchronized, to cultivate the conscious experience that lights up your world from the moment you awake. Cognition, then, is the operation that tirelessly decodes the world around you; a team of skilled personnel making sense of the flight information, scheduling your day's itineraries, and helping navigate the complex transit of daily life.

Nothing is lost in over-technical jargon here; instead, each explanation fits into the bigger picture, like the clockwork precision necessary for safe takeoffs and landings. The brain's processes are revealed step by step, each one critically dependent on the other, much like the collaboration required for an airport to function smoothly. The aim is not just to inform but to inspire: to share the marvel that is the human brain in a way that feels as real and tangible as the device you're reading this on, or the chair you're sitting in, reminding us that understanding our neural functions is understanding ourselves.

Dive into the grand terminal that is the brain, and you'll see how cognition mirrors the activity of a major airport. Here's a guide to this bustling hub, starting with sensory input – the arrival of passengers from various locations. This is akin to information received through the senses. Visual, audio, and other sensory data land on the runways of the brain's cortex, where they are greeted by a crew of neurons.

Next, this raw data is whisked away to be sorted and processed, much like luggage on conveyer belts. The thalamus takes on the role of the sorter, channeling sensory data to the appropriate terminals. Some input gets tagged as high-priority, like the sharp sound of a siren, which immediately gains the brain's attention – as emergency vehicles would on airport grounds.

At the heart of the operation sits the prefrontal cortex, the brain's air traffic control tower. It reviews the incoming information and decides on the next course of action – should the mental plane of thought be cleared for take-off, or should it remain grounded for further inspection? This part of the brain is where decisions are made, from the mundane choice of what clothes to wear, to the life-altering decisions that define futures.

Memory is another critical function, handled much like an airport's storage system for lost items. The hippocampus ensures that new experiences get efficiently cataloged and tucked away into long-term storage. When you recall a memory, it's pulled out and cross-referenced with new information, checked like a passport, confirming its accuracy before allowing it to inform your current situation.

Within this megastructure, neurotransmitters act as the service personnel, facilitating communication between different brain 'departments.' Dopamine might send a surge of motivation to the 'flight crew' of neurons, while serotonin works to maintain a stable 'customer service' mood throughout the brain.

But just as an airport has systems in place to ensure safety and efficiency, so too does the brain have mechanisms to balance and check its workings. Sleep is one such mechanism, acting as a mandatory downtime where the brain can perform maintenance – sorting through the day's accumulation of experiences and solidifying memories.

The grandeur and complexity of the airport are but a shadow of the vast capabilities of the brain, which not only processes countless pieces of data but generates the spark of consciousness – that which allows you to read, understand, and connect with these very words. It's this incredible process, this continual takeoff and landing of thoughts, that shapes the vibrant skies of your daily life.

Imagine the chief executive of a leading tech company, who sifts through vast data to steer decisions, much like a skilled chef judges which spices to mix into a perfect dish. That executive's brain is a whirlwind of cognition in action — evaluating options, predicting outcomes, and making choices that could impact thousands. Their prefrontal cortex is the busy kitchen of thought where strategies are cooked up and plans simmer on the backburner until they're ready to be served.

Now, picture an inventor in their workshop, surrounded by half-built dreams and the scent of soldering iron. Their task of solving complex problems and bringing ideas to life echoes the cognitive process of creative thinking — where imagination and practicality dance together like gears in a finely tuned clock. Each solution they craft is a testament to the brain's incredible ability to form new connections, like a bridge builder joining shores of insight across cerebral rivers.

In these everyday analogies, we can see the elegance of cognitive functions at play. Whether in the boardroom or the makerspace, the principles of cognition guide individuals as they navigate challenges, craft solutions, and shape the future. Each decision made, each problem solved, is a living example of cognition's might and influence, proving that the mind's inner workings are not just abstract concepts but are the very tools that forge reality.

Think of sensory memory as the pop-up notification on your phone screen — a fleeting glimpse at information that vanishes unless you engage with it. This is the immediate, yet temporary, capture of sensory information, like the brief image of a bird flying past the window, not yet saved to your mind's gallery.

Next up is short-term memory, akin to the device's RAM, where only a handful of apps can run at the same time. It's your mental workspace where you juggle a few tasks at hand before they're forgotten or saved into a more permanent form. It's where the brain processes and holds onto the day's grocery list until it's no longer needed — use it or lose it.

Long-term memory, on the other hand, is the vast and enduring hard drive, archiving experiences, skills, and knowledge. It's where the melody of a childhood lullaby, the grammar rules from school, and the wisdom from life's lessons are stored. Like files in a cloud service, these memories can be accessed across time and place, summoned into the workspace of short-term memory when needed to inform decisions or solve problems.

These types of memory form the infrastructure of cognition, much like the different data storage components of a computer work together to enable complex operations. They shape how we learn and behave — from the automatic muscle memory that allows a pianist to master a sonata to the recollection of a loved one's face — all made possible by the brain's remarkable capacity to store and retrieve a lifetime of moments. This voyage through the mechanics of memory illuminates how the very act of remembering is both a marvel and a cornerstone of the human experience.

Here is the breakdown on the nuanced process of memory in our brains:

- Sensory Input to Sensory Memory:

  - Primary sensory cortices:

- Involved regions: occipital lobe for vision, temporal lobe for hearing, etc.

　　　- Function: Act as the first receivers of external stimuli, interpreting raw sensory data.

　　- Neural firing:

　　　- Definition: A rapid electrochemical response to incoming stimuli.

　　　- Purpose: Creates a fleeting impression of an experience, akin to a snapshot.

　- Encoding into Short-Term Memory:

　　- Prefrontal cortex:

　　　- Role: Critical for selectively retaining and manipulating relevant information.

　　　- Involvement: Engages in tasks requiring focus and conscious thought.

　　- Repetition and attention:

　　　- Process: Reinforces neural pathways, boosting the chances of information retention.

　　　- Analogy: Similar to rehearsing a phone number so it doesn't fade away.

　- Short-Term to Long-Term Memory Consolidation:

- Hippocampus:

    - Importance: Essential for transitioning new, labile memories into stable, long-term ones.

    - Analogy: Like a librarian categorizing and shelving books for permanent storage.

    - Synaptic changes:

    - Mechanism: Long-term potentiation strengthens the connections between neurons.

    - Result: Solidifies experiences into memories that can last a lifetime.

  - Retrieval from Long-Term Storage:

    - Cues:

      - Types: Can be internal (thoughts) or external (environment).

      - Trigger: Act as keys that unlock the vault of long-term memories.

    - Memory networks:

      - Complexity: Involves a web of neurons spread across different areas of the brain.

      - Analogy: Like a search engine that retrieves files based on keywords.

Each part of this process is akin to a cog in the vast machinery of the mind, and understanding how they operate is paramount to appreciating the magic that is human memory. From the immediate reflex of turning your head at a flash of light to the profound recollection of a cherished childhood moment, the flow and interconnection of these stages underscore the complexity and miracle of how we remember. It's a meticulous system that underlies everything from academic learning to the feelings evoked by a familiar scent, proving that in the twilight of synapses, we find not only memory but the essence of our shared humanity.

Imagine the brain's decision-making process as cooking a meal. Just as you'd decide on a recipe, gather ingredients, and adjust spices along the way, your brain sifts through options, selects the most relevant information, and makes adjustments based on current inputs. It's a chef tasting the stew and deciding whether it needs more salt—a quick decision based on sensory perception and past cooking experiences.

Now, consider how you navigate rush hour traffic. Sensory perception in this scenario is like a driver behind the wheel, constantly scanning his surroundings—watching for brake lights, listening for horns, and feeling the car's movement. Each decision to speed up, slow down, or change lanes combines past driving lessons with split-second assessments of the cars' flow. Like this driver, cognition is always active, processing the input of the constantly changing road of life to arrive safely at your destination.

These everyday analogies peel back the curtain on the brain's cognitive processes, painting a picture of the hidden complexities behind seemingly simple tasks. Just as these activities are integral to daily living, so too is cognition fundamental to how we perceive, interact with, and make sense of the world around us. It is the unsung hero of routine life, orchestrating the symphony of actions and reactions that navigate us through each day.

Here is the breakdown on the cognitive processes that unfold in decision-making and sensory perception:

- Steps of Decision-Making:

  - Sensory Input:

- Process: Our senses act as scouts, gathering visual, auditory, and tactile information.

- Example: Seeing a red light and hearing a horn while driving.

- Thalamus:

- Function: Sorts the sensory information and sends it to the appropriate area in the brain.

- Example: Prioritizing the sight of the red light as critical information.

- Prefrontal Cortex:

- Task: Evaluates the information, considers the outcomes of various actions.

- Example: Deciding whether to stop abruptly or to proceed with caution.

- Basal Ganglia:

- Role: Pulls from past experiences to select the best action and initiates it.

- Example: Recalling similar traffic scenarios to execute the routine act of braking.

- Sensory Perception Process:

- Sensory Receptors:

- Function: Detect stimuli and translate them into nerve impulses.

- Example: Taste buds reacting to the bitterness of coffee.

- Primary Sensory Cortices:

  - Role: Begin the interpretation of these nerve impulses.

  - Example: Identifying that the bitter taste is indeed coffee.

- Association Areas:

  - Task: Integrate the information from various sensory cortices.

  - Example: Concluding that the coffee needs sugar based on taste and past preferences.

Each step in these processes is guided by the brain's previous experiences, which are stored as memories. A person's ability to react quickly to a red traffic light, for instance, relies on their past knowledge of traffic rules and the learned importance of the signal. Similarly, enjoying the morning coffee is not just about the taste—it's also informed by the memories and emotions associated with the ritual of drinking it.

By breaking down the cerebral choreography of decision-making and perception, it becomes clear how closely our routine actions and split-second choices are linked to a sophisticated neural interplay. It's a complex system made simple by the brain's incredible ability to adapt and learn, a testament to its unparalleled evolution and efficiency.

Picture a centuries-old painting hanging in a museum—each observer might see something different in the same brush strokes. Analogously, various philosophical theories of consciousness offer unique windows into the enigma that is human awareness. Materialism, for example, looks through a lens that sees consciousness as entirely a product of physical processes in the brain, akin to how one might view a painting merely as a composition of colors and textures.

Dualism, on the other hand, employs a lens that splits the image, offering one view of the physical brain and another of a non-material mind, suggesting a depth beyond the canvas that houses an essence invisible to the eye. These two theories stand in stark contrast, like comparing the work of an artist who strictly adheres to realism, to one who imbues their canvas with symbols and hidden meanings.

And then there's panpsychism, a lens tinted with the intriguing possibility that all matter could have a basic form of consciousness, reflecting a belief that the painting isn't just an object but has a presence and a story woven into every thread of its being.

Explaining consciousness is akin to trying to capture the essence of a masterpiece in a single phrase — it's a complex construct, where each theory frames it in a way that highlights different facets, contributing to our broader understanding. The exploration of these theories isn't about finding a definitive answer, but about appreciating the diverse and intricate viewpoints that echo through the art gallery of the human mind.

Let's take a closer look at the varied terrain of philosophical theories regarding consciousness:

- Materialism:

  - Core Assertion: Consciousness is a byproduct of physical processes in the brain.

  - Empirical Research: Studies using fMRI and PET scans show correlations between brain activity and conscious experiences, suggesting a material basis.

  - Neuroscientific Findings: Neuroplasticity and the effects of brain injuries provide a concrete map of how certain areas of the brain contribute to specific conscious functions.

  - Analogy: Like software running on hardware, consciousness emerges from the brain's complex neural circuitry.

- Dualism:

- Historical Perspective: Descartes' 'Cogito, ergo sum' introduces the concept of the mind existing separately from the body.

- Contemporary Views: Arguments revolve around the inability of physical science to fully explain subjective experiences and qualia - the individual instances of subjective, conscious experience.

- Subjective Experience: Personal feelings and thoughts might not be captured by examining brain activity alone.

- Intentionality: The mind's capacity to direct thoughts, suggesting a dimension beyond mere physical reactions.

- Analogy: Like a pilot and their aircraft, the non-physical mind directs the physical brain but is not reducible to it.

- Panpsychism:

- Core Idea: Consciousness is a fundamental feature of the universe, intrinsic to all matter.

- Philosophical Motivations: An effort to bridge the explanatory gap between physical processes and subjective experiences.

- Conceptual Implications: Introduces a paradigm in which simple forms of experiences are ubiquitous, with complex consciousness arising from complex material structures.

- Analogy: As every atom contributes to the shape of an object, every part of matter has a role in the tapestry of consciousness.

Each of these theories peels back a layer of the profound mystery of consciousness, offering distinct perspectives much like viewing a painting with different lights cast upon it. By dissecting and contrasting these arguments, we edge closer to grasping the elusive nature of our own awareness. Whether we see consciousness as an emergent phenomenon, a dual entity, or a universal trait, the exploration challenges and enriches our understanding of what it means to be conscious beings in a complex, interconnected universe.

Grasping scientific concepts can be like understanding how a smartphone app works. At first glance, it's a slick interface, but dive into the mechanics, and you discover a world of code and algorithms. For instance, take the concept of DNA replication—a complex process fundamental to life. Picture it as a meticulous scribe copying a precious manuscript, ensuring that each letter (base pair) is transcribed perfectly to preserve the text (genetic information) for future generations.

Consider, too, how neural networks in artificial intelligence are inspired by the human brain. These networks, like a vast series of interconnected roots in a forest, enable machines to learn from vast amounts of data. Each connection adjusts based on new information—similarly to how one's brain might strengthen a synapse each time a new skill is practiced, such as playing a chord on the guitar.

Moreover, exploring the physics of black holes, one might imagine a whirlpool in the ocean. Just as the whirlpool draws everything towards its center with the tug of its currents, a black hole warps space and time around it with its immense gravitational pull, captivating any matter that dares to venture too close.

Even quantum mechanics, often considered an exceptionally abstract field, has down-to-earth analogies. Quantum entanglement could be likened to a pair of perfectly synchronized clocks, no matter the distance between them—change one clock's time, and the other reflects that change instantaneously, defying the constraints of space.

Such connections between complex science and everyday phenomena not only make challenging concepts accessible; they reveal the poetry threaded through the fabric of scientific inquiry. It's a reminder that, at its core, science is a lens through which to witness and understand the marvels of the universe; it's a body of knowledge that is continuously evolving, offering new ways to see the world, much like discovering unexpected patterns in the stars every time we look up at the night sky.

Let's take a deeper look at the precise biological ballet that is DNA replication. Imagine the double helix of DNA as a zipper that slowly uncoils, courtesy of an enzyme called helicase, which splits the two strands apart much like you would untie a tightly knotted shoelace. Then enters DNA polymerase, akin to a skilled artisan, meticulously adding nucleotides to each original strand. This enzyme operates with both speed and precision, creating two identical DNA molecules from the original—each a mirror image of the other, ensuring the fidelity of genetic information for future cells.

Transitioning to neural networks, consider how an algorithm learns, shaped by vast reservoirs of data. Through a method much like a seasoned gardener pruning a tree, the algorithm is trained via backpropagation, trimming away the excess to optimize performance. Activation functions within this network serve as gatekeepers, determining which signals to pass through, paralleling how a musician chooses the right notes to play from a scale during a melody to evoke the desired harmony.

Now, peer into the enigma of black holes against the backdrop of Einstein's theory of general relativity. Visualize space as a stretchy fabric, with a black hole being a hefty object that creates a profound dip in this fabric. Then those daring enough to approach too close, like a marble spiraling towards a drain, experience the inescapable draw of this cosmic sinkhole.

Regarding quantum mechanics, let's ground superposition in something as everyday as a coin flip. Until you catch the coin, it is neither heads nor tails but in a blur of probabilities—analogous to quantum particles that exist in all possible states simultaneously until observed, revealing the randomness hidden within the universe's apparent order.

Exploring these marvels, from the microcosm of DNA to the vastness of cosmic phenomena, extends our appreciation for the intricate framework and profound beauty of science. It's the recognition of interconnectedness in all aspects of the universe, from the genetic scripts within us to the distant twinkle of stars, all bound by the same fundamental laws that govern existence. This journey through knowledge not only imparts a clearer sense of the world but also deepens our awe for the subtle complexities of nature and the human endeavor to decipher them.

As we close the book on this chapter of discovery, take a moment to marvel at the complexity and elegance wrapped within your mind. You've uncovered a world where synapses spark with the light of thought and neurons weave the fabric of consciousness, a place where the dance of cognition and sensation choreographs the very essence of experience. Think of each new insight as a further stepping stone into the rich terrain of your inner landscape—a chance to ponder the untold stories and unravel the enigmas that pulse in the quiet spaces between thoughts. May this exploration ignite a boundless curiosity, an enduring fascination with the brain's silent symphony. In turning the final page, usher in the beginning of a personal quest, a journey to delve further, understand more, and wonder—always wonder—at the endless marvels of the human mind. Through the simple act of learning, you've equipped yourself with a finely-tuned lens to better view yourself and the universe. Stand confident in this newfound knowledge, and carry it with you as a torch that lights the path of inquiry and reflection.

EMOTIONS AND THE SOCIAL BRAIN

Dive into a world as intricate and fascinating as the inner workings of a clock, one that ticks and tocks to the rhythm of human connection—welcome to 'Emotions and the Social Brain'. This chapter is akin to lifting the face of that clock to reveal the gears and springs beneath. Here, you'll unravel how the neural circuits, the hidden wires and switches of our brains, are key players in the dance of social interaction.

Imagine each feeling you have as a spark that travels along these circuits, connecting with others like a phone call links two distant friends. These emotional signals guide how you relate to people around you—much as a handshake or a smile opens a conversation. They are a silent language, shaping everything from a mother's affectionate gaze to a leader's rousing speech.

The aim here isn't to simply parade scientific facts but to offer you a lens to clearly see the marvels behind every laugh, every tear, and every heartwarming reunion. By breaking down scientific concepts into their basic parts, like a storyteller unravels a tale bit by bit, we'll navigate from the shores of neurobiology to the vast oceans of human relationships. Equipped with this knowledge, you'll not just understand the "what" and "how" of our social minds, but also peer into the profound "why" behind our shared smiles and silent supports.

Picture your emotions as a complex network of rivers weaving through the landscape of your brain. These waterways are bustling with activity, directed by a series of dams and channels—the brain regions and neurotransmitters—that orchestrate the flow of your feelings. At the heart of this system sits the amygdala, like a watchtower, scanning for emotional significance and sounding alarms when necessary. The prefrontal cortex acts as a seasoned captain, steering the emotional currents, deciding when to let the river run wild and when to tighten the sluices for control.

Dopamine and serotonin flow through these channels as chemical couriers, carrying messages that alter the mood and affect of the mind.

Picture dopamine as the current that propels you to reach for rewards, while serotonin works to balance the waters, keeping the emotional ecosystem stable. Together, these neurotransmitters, alongside others like oxytocin—the molecule of connection—form a delicate balance that dictates how we experience joy, fear, love, and sorrow.

This intricate description doesn't just lay out the 'what' of emotions but deepens your grasp of the 'how'. Understanding these neural underpinnings isn't about memorizing terms; it's about realizing that every burst of laughter or pang of sadness has a place in the vast map of your brain. It's a recognition that our most profound human experiences are also the most beautifully complex, and to understand them is to inch closer to the essence of what it means to live, love, and interact with the world around us.

When an event stirs an emotion within you, think of it as a stone thrown into the still waters of a lake, creating ripples. This is the initial trigger, an event or thought that catches your attention and activates your emotional response. Immediately, your brain's amygdala, much like a vigilant sentry, evaluates this disturbance, determining its importance, and setting the stage for how you will emotionally react to it. Is it a threat? Is it a joyous surprise? The amygdala reads the cues and calls the shots.

Neurotransmitters then enter the mix, adding depth and color to the emerging emotional response. Dopamine surges in response to pleasure or the anticipation of a reward—like sweetening a dessert to enhance its flavor. Serotonin flows in to stabilize your mood, ensuring that the emotional response is appropriate—comparable to a musician moderating the volume of their instrument to find balance within a symphony. Meanwhile, oxytocin comes into play during social interactions, fostering feelings of trust and bond—it's the warmth of a hug in chemical form.

As these neurotransmitters course through your neural pathways, the prefrontal cortex steps in like an expert conductor, managing the intensity and complexity of the emotional experience. This regulatory powerhouse can dial down the heat of anger or fine-tune the excitement of anticipation, maintaining equilibrium as deftly as a thermostat keeps a room's temperature steady.

Finally, this sophisticated emotional circuitry plays a pivotal role in social communications. The nuanced interplay of brain regions and chemicals allows us to perceive and share feelings with others—such as understanding a friend's sadness or experiencing the contagious joy of a crowd. This system is the bedrock of empathy and connection, enabling us to navigate and flourish within our social worlds with the subtlety and sophistication that define our humanity.

Imagine your brain's emotion-processing network as a bustling train station, where trains represent your feelings, dispatching to various destinations that are your reactions and interactions. The network's train tracks, much like the pathways of neural communication, wind through cities and towns—these are the social contexts where we live our lives. Just as train conductors must communicate and coordinate to ensure passengers reach their loved ones, the molecules and signals within our brain need to sync up to guide us through the social landscape.

Our emotional expressions serve as tickets, invitations to others to join us on our journey. When someone shares a story with you, their emotions invite you onto their train, asking for companionship on their route. If your tickets match, you find common ground, and just like that, a connection is made. Similarly, observing someone else's emotional train can evoke empathy, and we might adjust our own schedules—our behavior—to accommodate.

The amygdala is the station master, vigilant and always assessing, ensuring safe conduct and correct emotional responses; while our neurotransmitters are the signals along the tracks, changing the points to direct this intricate web of trains. The prefrontal cortex, then, is the grand central switchboard, keeping the whole system running smoothly, deciding when to put on the brakes or let a feeling accelerate.

This vivid tapestry of movement and exchange is the very essence of social relationships, with each shared smile or comforting embrace akin to a rendezvous of spirits on the platform. It is in this dance of arriving and departing, of signaling and responding, that our bonds are formed, maintained, and celebrated.

Here is the breakdown on the intricate relationship between emotions and social behavior, and the neurobiological processes that underpin them:

- Triggering of Emotions by External Events:

  - Input:

    - Environmental stimulus: Any external situation perceived through the senses, like a friend's smile or a sudden loud noise.

- Sensory data: The information collected by the senses, relayed to the brain for processing.

- Processing:

- Amygdala: This almond-shaped structure evaluates the emotional relevance of sensory data.

- Emotional response initiation: If a threat is perceived or a positive stimulus recognized, the amygdala quickly triggers a corresponding emotional reaction.

- Role of Neurotransmitters in Shaping Emotional Experience:

- Dopamine:

- Links to motivation: Spiking during pleasurable activities and encouraging repeating rewarding actions.

- Influence on social behavior: Can drive the pursuit of socially rewarding experiences, like bonding with friends.

- Serotonin:

- Mood regulation: Known for leveling mood swings and promoting feelings of well-being.

- Social behavior modulation: Helps to temper aggressiveness and encourage social cooperation.

- Oxytocin:

  - Social bonding: Often called the 'love hormone', plays a crucial role in building trust and strengthening social ties.

  - Empathy and relationships: Facilitates the understanding of others' emotions and fosters deep connections.

- Function of the Prefrontal Cortex in Emotional Regulation:

  - Higher-order processing:

    - Consideration of context: Where the brain weighs current emotions against past experiences and future consequences.

    - Rational assessment: The process of making sense of emotions and deciding on appropriate responses.

  - Response modulation:

    - Braking system: Just like pressing a car's brake, it suppresses impulsive reactions.

    - Fine-tuning responses: Adjusting reactions to fit socially acceptable norms and personal goals.

- Impact of Neurobiological Processes on Social Interactions:

  - Social cues:

    - Reception: Recognizing facial expressions, tone of voice, and body language.

- Interpretation: Understanding what these cues mean in a social context and deciding how to respond.

- Empathy:

- Emotional mirroring: The ability to feel what others are feeling as if you were in their shoes.

- Compassionate response: Offering comfort or solidarity based on shared emotional understanding.

- Bonding:

- Connection initiation: Like the first thread woven between two friends.

- Relationship strengthening: Deepening trust and rapport over time through shared experiences and mutual empathy.

Understanding the flow of emotions through our brain's complex circuitry and the resulting symphony of social interaction can offer profound insight into the nature of our relationships. Every smile shared, every hand extended in comfort, becomes a meaningful exchange in the rich commerce of human emotion and connection.

Consider the immediate quiet that blankets a room when someone enters in tears; it's a visceral response to a display of sadness that resonates with a fundamental part of our humanity. Everyone in the space might feel a tug in their chest, an empathetic reflection of distress not their own, prompted by mirror neurons that quite literally allow them to feel a semblance of that sorrow. The interaction here is subtle yet powerful—people might offer tissues, a comforting touch, or a sympathetic ear without a second thought. This is emotional expression and regulation occurring in real-time; the group's collective behavior molds to accommodate and console.

In contrast, envision a boardroom when a leader, fueled by a cocktail of adrenaline and cortisol during a stressful situation, exudes confidence and takes charge. The team, previously a sea of uncertainty, now follows suit. The emotional regulation here, likely subconscious and honed by years of experience, infects the group's dynamic, dictating behaviors that pivot from anxiety to assertive action.

These instances shine a spotlight on the silent, sometimes invisible threads that draw our reactions and decisions—from a stranger's kind smile that lifts your mood and inadvertently leads you to hold the door for someone else, to the infectious enthusiasm of a coach that propels a team to victory. It's the recognition that the ebbs and flows of our inner emotional lives are far from private escapades; they forge the foundations of every human interaction, shaping individual paths and the broader social tapestry with every heartbeat and breath.

Let's take a deeper look at how the ripples of one's emotions can send waves through the entire social landscape. As an emotional stimulus arises—be it a touching scene in a movie or the tense atmosphere of a negotiation—our sensory organs are the first to pick up on it, much like antennae tuned to the world's broadcasts. This sensory information lands in the lap of the amygdala, that ever-vigilant gatekeeper that quickly thumbs through memory's archives to give context to what we're sensing. Is this something to fear? To cherish? The amygdala, with lightning-fast appraisal, sets off a cascade.

Cue the mirror neurons, those empathetic mimics nestled in the brain, firing in sympathy when we witness someone else's emotion. Watching a friend's face crumple in sadness can spark a similar pattern inside our own minds, a silent chorus of neurons singing the same mournful tune. It is here in this neural mimicry we find the roots of empathy, a shared emotional beat that syncs us with our companions.

Meanwhile, like conductors of an invisible orchestra, our bodies respond to the stress of a challenge or the thrill of triumph with a surge of hormones. Cortisol and adrenaline are released, priming the body for action—a leader's firm stance or raised voice might be the upshot of an adrenal tide, signaling resolve to the group. This show of confidence can spread through the collective, transforming a room's energy from hesitant whispers to a unified march.

This is no one-way street; the feedback loop comes into play as a circle of validation and reinforcement. The team's responsive nodding and rallying cries bolster the leader's outward poise, an emotional echo chamber wherein each reflection amplifies the original sentiment. The result is a dynamic emotional ecosystem, navigating the ebbs and flows of human interaction, where every raised eyebrow or clasped hand is a verse in the grand social symphony. Understanding these neurobiological undercurrents not only shines a light on the dance of connection but empowers us to be more mindful participants in the intricate ballet of the social brain.

Oprah Winfrey, Nelson Mandela, and Elon Musk – three paragons of leadership, each employs the social brain in compelling yet distinct ways.

Oprah Winfrey exemplifies emotional intelligence, connecting with audiences on a deeply personal level. Picture her as a conductor of a vast orchestra, every interview a concerto of shared stories and feelings. Her ability to attune to her guests and empathize allows for a concert where the melodies of individual experiences resonate with the collective. Her emotions guide the tempo, her sincerity sets the rhythm, and her audience feels every note, creating a harmonious symphony that rings with authenticity and compassion.

Nelson Mandela, on the other hand, wielded the power of emotional fortitude and reconciliation—his heart, a lighthouse guiding a nation through the fog of discord to the shores of harmony. His speeches often served as a beacon, his tone and poise uniting warring factions not only through words but also through the quiet determination and hope that underscored them. His leadership was a masterclass in emotional regulation, embodying the resilience and understanding needed to inspire change and foster unity.

Elon Musk demonstrates the impact of visionary enthusiasm coupled with the calculated risk of unpredictable emotionality. Envision him as the maverick pilot of a spacecraft, his Twitter communiqués steering public sentiment and market waves alike. The passion in his vision acts as a gravitational force, drawing in supporters and investors, aligning their energies with the trajectory of his ideas. His emotional expressions, delivered in bursts of tweets, can act as thrusters propelling his ventures forward or turbulence shaking stakeholder confidence.

These leaders show that in the realm of public influence, the social brain is your strongest ally—the compass in navigating the intricate web of human connections, the map charting the course of collective emotion. It's the understanding that while the science of emotion is universal, its art lies in the way we harness it to forge bonds, lead, and inspire.

As we draw the curtains on this exploration of the social brain, let's pause to appreciate its silent orchestration of our emotional lives. Our everyday interactions, the tapestry of friendships, enmities, and love, are all directed by this unseen maestro within our skulls. Its influence is profound, touching everything from the empathy we extend to a stranger to the ties that bind families and communities. The science of this is intricate, but the evidence of it plays out in the simplest of our actions—the helping hand offered instinctively or the shared laughter that forges an instant bond.

Reflect on moments in your life when a glance or a word has shifted the mood of an entire room—such is the power of the social brain. It's a tool for connection as innate as breathing, honed through evolution to ensure that no one goes through life's odyssey alone. This understanding isn't just academic; it's a mirror reflecting the deep connections that make up your world.

By grasping the significance of the social brain, you're equipped with a compass to better navigate the complex social seas, to understand their currents and climate changes. Harnessing this knowledge in daily life illuminates the intricate dance of human connection, empowering us to choreograph our steps with greater awareness and sensitivity. The hope is, with this knowledge in hand, you stride into your world more attuned to the emotional ebbs and flows that color every interaction, and use it to enrich not just your experiences but those of the people around you.

NEUROPLASTICITY THE ADAPTABLE BRAIN

Now we peer within the doorway and into the ever-changing world of the human brain, where the very fabric of thought, memory, and experience continuously weaves itself into new patterns. Imagine if the mind were a home—its rooms representing various brain regions, and its habitants, the neural connections that live there. Just as a family might renovate their abode to accommodate a new member, or repurpose a space after its old role has faded, the brain alters its internal architecture in the face of new learning or in recovery from injury.

This phenomenon, known as neuroplasticity, is akin to a city that's forever under construction, building new bridges of understanding or repaving worn paths of habit. It's a personal journey, where one's dedicated practice at the piano or perseverance in learning a new language can physically shape the landscape of the brain, forging connections as tangible as the streets under our feet.

Begin this exploration with both curiosity and wonder, knowing that the brain's potential to remodel itself is as natural as it is extraordinary. Along the way, you'll discover the magnificent agility with which the brain adapts—how it can reforge broken links, carve new pathways, and even reassign functions, ensuring the resilience of the mind. Each page turned and fact learned on this journey will not only illuminate how adaptable the brain is but also offer a reflection on the adaptability within oneself.

Neuroplasticity is the brain's extraordinary ability to reorganize itself. Picture a vast network of roads in your mind; these are your neural pathways, the routes that information travels along. Just as a city evolves, building overpasses and new highways, your brain modifies these paths in the face of fresh experiences and knowledge acquired.

When you learn something new, your brain is not just absorbing information—it's also constructing new connections. These are physical changes, like new branches sprouting from a tree, extending out to join with other branches. The more you use these new pathways, the stronger and more established they become, making it easier to access that information or skill—it's a bit like walking a path through a field; the more it's traveled, the more defined it gets.

Meanwhile, when injuries occur, the brain can sometimes reroute these roads of thought, tapping into the concept of neuroplasticity to transfer functions from a damaged area to a healthy one. It is as if when one road closes down, the brain's GPS immediately seeks alternate routes to ensure the journey can continue.

Understanding neuroplasticity gives us clarity about how habits are formed, how skills are mastered, and how, even after damage, the brain can recuperate and adapt. With each new experience, you're not just learning—you're physically reshaping your brain's structure, a testament to our inherent capacity for change and growth.

Let's take a deeper look at the microscopic details that make neuroplasticity a cornerstone of the brain's adaptability. Every new experience or recovery from trauma engages a delicate dance at the cellular level within the brain's complex lattice. Synapses, those tiny gaps between neurons, act as bustling intersections where neurotransmitters—brain's chemical messengers—leap across to relay signals. The receiving dendrites, tree-like extensions from a neuron, pick up these signals with receptor sites that are remarkably adaptable, increasing or decreasing their sensitivity based on activity levels.

When learning, synapses might bolster their connectivity, a process called synaptic plasticity. Picture this as shaking hands with someone; the more frequently you shake hands, the stronger the grip becomes. Similarly, neurons fortify their connections through repeated engagement—this is the essence of long-term potentiation, a process that underpins how learning embeds itself as memory. It involves an increase in neurotransmitter release, a rise in receptor efficiency, and growth of new dendritic spines—think of it as adding additional phones for a conference call to ensure the message gets across clearly and memorably.

Neurogenesis, the formation of new neurons, unfolds most conspicuously in the hippocampus, the brain region associated with learning and memory. This is like planting new seedlings in a garden, which grow and intertwine with established plants to enrich the ecosystem.

And then there's recovery. After injury, the brain's plasticity allows it to reroute functions from injured areas to healthy ones—neuronal rerouting is akin to detours set up after a road has been blocked. Cortical remapping takes place, reallocating tasks much like a team taking on the responsibilities of an absent colleague. Glial cells, once thought to merely be the brain's support cells, play a pivotal role in this process, offering structural support, cleaning debris, and modulating inflammation to create an environment conducive to repair and reorganization.

Understanding these intricate cellular processes paints a picture of a brain that is never static, constantly rewiring and adapting—mirroring the never-ending change in our lives. This knowledge isn't just a scientific detail; it's a reflection of human resilience and the potential for transformation and healing, underscoring the profound adaptability innate in every individual.

Consider the brain's ability to learn as if it were a network of pathways in a vast forest. Each time you acquire a new skill or piece of knowledge, it's like cutting a fresh path through the underbrush. At first, the route is rough and obscured; traveling it requires effort and attention. However, with each repetition, with every revisit, the path becomes more defined, the undergrowth trampled down, until it's a well-worn trail that you can traverse with ease.

These neural pathways, much like forest trails, are reinforced and strengthened through use. Just as a frequently walked path becomes easier to follow over time, so too do the connections in our brain solidify with repeated activation—a testament to the old adage, 'practice makes perfect.' The more you practice your French vocabulary or piano scales, the stronger these neural 'paths' become, facilitating smoother and quicker recall akin to a jogger's sprint down a familiar track.

And let's not forget those pathways that fall out of use. Just like a neglected trail which slowly succumbs to the creeping vines and encroaching trees, neural connections can weaken and fade over time if they're not engaged—a reminder that our brains, much like the trails, thrive on exploration and regular travel. It's in this elegant interplay of use and disuse that the brilliance—and the boundaries—of our capacity to learn and remember is vividly illustrated.

Here is the breakdown of how the flow of new information takes root and grows within the brain, transforming into a robust part of one's neural architecture:

- **Initial Neural Response to New Information**:

    - **Synaptic firing**: Neurons communicate by sending electrical impulses down their axons, reaching out to the next neuron like a lightning bolt jumping across the sky.

    - **Neurotransmitter release**: At the synapse, chemicals such as glutamate pass the message along, acting like a ferry carrying cars across a river, bridging the gap to the next neuron.

- **Strengthening New Neural Pathways**:

    - **Synaptic efficacy**: With practice, the synaptic transmission becomes smoother and more efficient, akin to a well-oiled machine running at top performance.

    - **Structural changes**: Dendritic spines, the tiny outgrowths on neurons, proliferate and reform, much like branches reaching out to sunlight, forming more synapses and solidifying the route of transmission.

- **Consolidation Phase: Short-Term to Long-Term Memory**:

    - **Neural network integration**: New bits of information intertwine with established cognitive webbing, like interlocking threads in a tapestry that provide context and meaning.

    - **Role of sleep**: During rest, the brain's cleaning crew comes in, solidifying these connections, a process similar to saving a document to a hard drive—transferring data from the temporary cache to long-term storage.

- **Weakening of Unused Pathways**:

    - **Synaptic pruning**: Just as gardeners trim back overgrowth to invest more energy into flourishing plants, the brain selectively weakens lesser-used connections, preserving resources and efficiency.

This meticulous sequence from the spark of learning to the concretization of memory displays how our brains adapt to the ever-changing landscape of experience. Understanding these steps is like holding a roadmap to the cognitive terrain, revealing not only the marvels of the mind's capabilities but also inspiring a deeper appreciation for the daily acts of learning and remembering.

In the remarkable journey of recovery after a brain injury, neuroplasticity plays a role as pivotal as a lead actor on stage. It's the brain's inbuilt resilience factor, an ability to stitch itself anew, reorganizing functions and pathways in response to damage. This process might sound like a mystery, but it unfolds with a logic as comprehensible as it is fascinating.

When a brain area is injured, untouched regions take on new roles, a bit like neighbors helping out when someone's house is in disrepair. These regions can develop new abilities or strengthen existing ones to compensate for lost functions – think of it as taking up a new hobby or skill when your usual pastime is no longer an option. This is neuroplasticity in action: neurons reaching out across the void left by injury, building new connections.

Specialized brain exercises and therapies can enhance this recovery, plotting a course for the brain to rewire itself. The practice is akin to rerouting traffic after a main road's closure – it's challenging and slow at first, but with dedication, alternate paths become as smooth as the original was. Here, the brain demonstrates its remarkable ability to not just heal but to reinvent its pathways, offering hope where loss once took center stage.

This detailed account showcases not just the process but the profound potential for healing and adaptation within every brain. It reflects a story of resilience and regeneration, underscoring the significance of neuroplasticity in not just surviving, but thriving, following the trials of injury.

After a brain injury, the pathway to recovery is a multifaceted process, facilitated by the brain's natural ability to adapt—its neuroplasticity. Here's a step-by-step guide to understanding how this happens:

Stages of Injury and Recovery:

- **Acute Response**: In the immediate aftermath of injury, the brain reacts swiftly. Neurons surrounding the injured area can dampen their activity as a protective measure, while the immune system dispatches cells to clean up debris. This is the cerebral equivalent of emergency responders at a scene, working to prevent further damage.

- **Subacute Phase**: Subsequent weeks bring the early recovery mechanisms online. Neurons that weren't destroyed start to rewire their connections. It's as if the brain, encountering a roadblock, begins forging detours around the closed-off route.

- **Chronic Phase**: Over months and years, these makeshift paths can become more permanent, and rehabilitation efforts focus on reinforcing and refining them. It's the long haul of reconstruction after the initial crisis has passed.

Cellular Repair Processes:

- **Glial Response**: Astrocytes and microglia—the brain's support and immune cells—mobilize to repair and form a scar in the brain tissue. Like a construction crew, they clear the rubble and lay down new groundwork, providing structure for new growth.

- **Axonal Sprouting**: Neurons begin to extend new projections, axonal sprouts, to make up for lost connections. Picture a plant growing new stems to find sunlight; similarly, these sprouts reach out to reestablish communication with other neurons.

Therapies That Harness Neuroplasticity:

- **Cognitive Rehabilitation**: Custom exercises gear up to mentally challenge the brain, prompting fortification of cognitive functions, akin to adding weight during strength training.

- **Motor Relearning**: Gradual and repetitive movement exercises advocating for the re-establishment of patterns. Like learning to dance by following footprints on the floor, the brain and body relearn to move in tandem.

This guide lays out a map of how the brain traverses the terrain of injury and repair. Through the intricate process of neuroplasticity, cellular-level adjustments correspond with behavioral therapies, all contributing to the ultimate goal of recovery and a return to function. As readers venture through each phase and process, they gain an appreciation for the complexity and capability of the brain to reroute, rebuild, and recover, translating a potentially overwhelming topic into relatable and manageable segments.

Dive into the world of neuroplasticity, and you'll find it's not just a concept tucked away in neuroscience labs – it's present in transformations witnessed across the public stage. Take Gabby Giffords, the U.S. Congresswoman whose recovery from a brain injury became a beacon of hope. Her journey parallels the incredible way our neural pathways can rebuild, much like a city's roads after an earthquake. Effortful at first, her speech and movement therapies gradually carved out new neural pathways, akin to trails being blazed in a dense forest, guiding her towards remarkable recovery.

In the arts, consider guitarist Pat Martino, who relearned his lost skills after surgery for a brain aneurysm. The silent melodies in his mind found new expression as his fingers, following the invisible scores imprinted in his resilient brain, danced once again along the strings. His renewed virtuosity sings of the power of practice and persistence, the key drivers of neuroplastic change, illuminating a path that turned adversity into a harmonic resurgence of talent.

Neuroplasticity shows itself not solely in recovery but in everyday feats as well. Elon Musk's ventures into space and electric cars speak to the brain's ability to foster innovation and adapt thinking to changing horizons. Each entrepreneurial pivot can be seen as a neural shift, a cognitive recalibration to the demands of fresh challenges and the allure of uncharted territories.

These stories, among others, offer tangible, stirring examples of how our brains possess a remarkable, almost magical malleability that's responsive to the demands we place on it, whether in relearning a skill after trauma or pioneering paths no one has tread before.

In the realm of neuroplasticity research, scientists stand on the brink of a landscape vast with possibility, yet navigate fields strewn with challenges. The central struggle lies in mapping the elusive nature of the brain's adaptability — it's like trying to chart the ocean's currents while they are in constant, unpredictable motion. The intricacies of individual differences add layers of complexity; each brain wears the grooves of its unique journey, and what aids one mind's adaptability may not suit another, just as a key tailored to one lock cannot open all doors.

Current research grapples with translating findings from the controlled setting of the lab, where variables can be contained and carefully manipulated, to the chaotic theater of real life, where variables run wild. Think of it as moving from rehearsing a play on a quiet, empty stage to performing on opening night with unforeseen hiccups in front of a full house.

Limitations also rest on the boundaries of neuroplasticity itself. There is no carte blanche promise of healing; it's not a panacea for all neurological ailments. The brain has a remarkable capacity to adapt, but this capacity is not infinite. It's as though our brains have a budget of resources, and while they can juggle and reallocate funds to a remarkable extent, the coffers are not bottomless. The recovery and reorganization potential may be tempered by age, the extent of injury, or even genetic predispositions, with each element influencing the neuroplastic machinery's efficiency.

Faced with these challenges, researchers and clinicians continue to press forward, their endeavors carving out new understandings and therapies that inch closer to maximizing the human brain's adaptive potential. The journey to unravel neuroplasticity's enigmas is marked with the diligence of a sculptor — each chisel stroke reveals more of the figure hidden within the marble, and each study uncovers more about the brain's capacity to rewire and heal.

Let's take a deeper look at the contours and shades of neuroplasticity's limits. While the brain's ability to form new neural connections is remarkable, it doesn't operate in an open field without fences.

Beginning at the cellular level, neurogenesis—the birth of neurons—and synaptic growth aren't endless. Just like a tree reaches a point where it no longer sprouts new limbs, the brain has its own checkpoints. Each neuron goes through a life cycle, and there are only so many rounds of this cycle it can complete, dictated by the availability of crucial growth factors—similar to how a plant needs water and sunlight to thrive.

When it comes to age, imagine the brain as putty: young brains, like fresh putty, are malleable and quick to shape. In contrast, older brains have set and require more effort to remold. This isn't just metaphorical; scientific evidence suggests that a young brain's plasticity is more pronounced, with greater synaptic malleability and recovery rate post-injury.

The extent and exact site of an injury also play starring roles in neuroplasticity's effectiveness. A minor scratch on a digital device's screen might not hinder function, but a crack over the camera's lens affects performance severely. Similarly, the size and location of damage in the brain can limit how effectively neuroplasticity can compensate or redirect functions.

Lastly, genetic predispositions underpin much of how plastic a brain can be. Genetics are akin to having a blueprint or a set of assembly instructions. Some individuals might come with a manual that supports robust neural rewiring, while others work with pages that make these constructions more challenging.

Hand in hand with continuous research, understanding these boundaries helps set realistic expectations and optimizes rehabilitation strategies. It pinpoints where the brain can push through obstacles and where additional support might be needed, paving a road for therapeutic advancement and fine-tuned interventions. This insight is about recognizing the power of the brain's adaptability while understanding that like everything in nature, it operates within a framework of possibilities and constraints.

Neuroplasticity stands as one of the brain's most fascinating and essential qualities, the very mechanism that sculpts our experiences and molds our identities from infancy to adulthood. It's the hidden artisan behind the vast gallery of human capacity, from the simple act of remembering a phone number to the complex undertaking of learning a new language or recuperating from a stroke.

With every challenge we encounter, every new skill we take on, the brain flexibly adjusts, rewires, and strengthens its neural networks—reinforcing the reality that we are constantly evolving, learning beings. This organic rewiring isn't just a response to learning; it's a testament to our resilience, the silent strength that enables someone to dance again after an injury or reshapes a life after a significant change.

As research persists, the horizon of neuroplasticity stretches even wider. Scientists are only beginning to unlock the potential therapeutic uses of this knowledge. Future works may reveal ways to harness this inherent adaptability to help heal cognitive impairments, aid recovery from neurological damage, and perhaps even delay the effects of aging. These pursuits could redefine the limits we've placed on learning, memory, and brain rehabilitation, offering hope for an increased quality of life to countless individuals.

So, as we gaze into the future, poised at the edge of discovery, the brain's adaptability reminds us that change is not only possible, it is etched into the very cores of who we are. This dynamic plasticity reverberates through every aspect of our lives, empowering both the growth of each person and the progression of humanity at large.

CONCLUSION

As we close the final chapter of "Neuroscience Made Easy," it's time to look back on the odyssey we've embarked on together, exploring the tantalizing complexities of the human brain. From the firing of neurons to the dance of neurotransmitters, each concept we've unearthed has brought us closer to understanding the inner workings of our minds.

We've journeyed through the landscape of the brain—discovering how memories are imprinted like footprints on a sandy beach, untangling the circuitries of emotion as elaborate as city skylines, and exploring consciousness akin to gazing into the vast expanse of the cosmos. Throughout these pages, neuroscience has become less of an arcane field and more like a familiar map, one where every reader can find their way.

The key themes we've encountered—the brain's malleability, the interplay between brain structure and function, the role of genetics and environment, the potential for healing—aren't just abstract ideas. They're the very fibers woven into the tapestry of daily living. We've learned that the brain's adaptability can be harnessed, that habits can be reshaped, and that no part of our personal growth is beyond the reach of change.

Reflecting on the impact of this book, it's clear that neuroscience isn't just for academics or practitioners. It's a field with the power to inform and transform every one of us. The insights gleaned here have the potential to ripple out into how we learn, communicate, and care for our mental health—compass points leading us toward more mindful and enriched lives.

As you step beyond the final page, consider how the brain's boundless faculties might illuminate your own path. Remember that every chapter of knowledge opens new doors—doors to self-discovery, to greater empathy, to the uncharted territories of the mind. The dialogue between neuroscience and everyday experience doesn't end here; it's an ongoing conversation, one that will continue to evolve, much like the brain itself.

Thank you for turning these pages and allowing "Neuroscience Made Easy" to guide you through the complex, yet infinitely accessible world of brain science. May the journey through the labyrinth of neurons and synapses leave you with a lifelong fascination for the marvels of the mind.

ABOUT THE AUTHOR

Jon Adams is a Prompt Engineer for Green Mountain Computing specializing and focusing on helping businesses to become more efficient within their own processes and pro-active automation.

Jon@GreenMountainComputing.com

Printed in Great Britain
by Amazon